# WELL-SITE GEOLOGIST'S HANDBOOK

# WELL-SITE GEOLOGIST'S HANDBOOK

**DONALD McPHATER**

**BRIAN MacTIERNAN**

*McKinlay-Smith Int.'l Ltd.*

PennWell Books
PennWell Publishing Company
Tulsa, Oklahoma

**PennWell Publishing Company**
1421 South Sheridan Road/P. O. Box 1260
Tulsa, Oklahoma 74101

**Library of Congress Cataloging in Publication Data**

McPhater, Donald.
Well-site geologist's handbook.

1. Petroleum—Geology—Handbooks, manuals, etc.
2. Oil wells—Handbooks, manuals, etc. I. MacTiernan, Brian. II. Title.
TN870.5.M39 1983 622'.338 82-22249
ISBN 0-87814-217-7



Printed in the United States of America

1 2 3 4 5 87 86 85 84 83

# CONTENTS

# INTRODUCTION

Retrieving information from the well site is subject to many factors, including personnel experience and diligence, human error, lack of communication, adherence to standard valid procedures, and physical conditions. The experienced well-site geologist can glean the maximum amount of information from well-site data and put a realistic statistical bias on its likelihood of being correct. Conversely, without personal experience of well-site geology, one is not fully qualified to interpret well-site information. For this reason, well-site geology should not be relegated to the realms of pure sample catching and description.

The *Well-Site Geologist's Handbook* is a practical well-site geological package that is pitched at a level that can be attained practicably at the well site. It is designed to be of a size and depth of detail that is usable, not stifling. Thus, it assumes a basic geological background and does not aim at being a standard reference.

This book is aimed primarily at the practicing well-site geologist, for whom it is hoped that the manual will be a valuable checklist on the procedures involved. Second, the book may be used as a basis for training those who are about to be involved in well-site geology. Third, it is hoped that this book will whet the appetite of geologists who are as yet undecided upon a line of career, and that it will give them some insight into well-site geology—which, of course, forms only a small part of hydrocarbon exploration as a whole.

# 1. PREPARATION

## 1.1 Materials and Equipment

### 1.11 Sample Study

A typical list of materials and equipment required to carry out a fairly comprehensive practical well-site evaluation follows. The list obviously is subject to individual and operator requirements.

—6–30-diameter binocular microscope
—Light source of microscope. A Burton lamp or any source producing minimal color distortion is recommended.
—Watch glasses and spot plates. These should be glass because plastic's surface tension is too high.
—Window glasses for silt content evaluation
—Test tubes and stoppers for oil stain test
—Eye droppers
—Tweezers and picks
—Chemicals

The list of chemicals depends on the depth of investigation. Care should always be taken when handling chemicals, especially carcinogenic ones.

—Strong HCl
—Dilute HCl
—Stain test for carbonates. This solution consists of 0.2% HCl, 0.2% alizarin red S, and 0.5% saturated K ferricyanide.
—Bromoform for distinction between gypsum and anhydrite
—Phenolphthalein to identify cement
—Oil solvent, such as chlorothene

—Log and report sheets; drafting materials
—Company standard legend
—Munsell color charts
—Ultraviolet lamp

## 1.12 Core Packaging

The above list is required when studying core samples. Following is a list of materials and equipment to package the core safely and efficiently.

—Core boxes and nails. Some operators use intermediate core trays for temporary storage when the core comes out of the barrel. If possible, this should be avoided since it increases the chances of error.
—Plastic lay-flat tubing and heat sealing machine, if required
—Plastic bags for broken cores
—Transparent wrap such as cling film
—Aluminum foil
—Wax and wax bath or other core sealant
—Staple guns and staples
—Nylon-reinforced tape
—Oil- and water-proof marker pens plus solvent to revitalize clogged pens
—Measuring tape
—Geological hammer
—Cutting knife
—Report sheets and note pad
—Padding materials, such as shredded paper, loose rags, or foam

Again, individual or operator requirements may change the above list.

## 1.13 Other Materials

The well-site geologist may be required to ensure adequate materials and equipment for other specialized services, e.g., for source rock analysis or micropaleontological studies.

### 1.14 Personal Gear

The well-site geologist should be sure that his personal gear meets the required safety standards and ensures his comfort in expected environmental conditions.

## 1.2 Readiness of Technical Services

Mud-logging and associated personnel, i.e., data acquisition and pressure engineers, should be aware of their duties and responsibilities to the well-site geologist and engineering staff regarding sampling intervals and reporting procedures. The well-site geologist should also check that these personnel have adequate supplies.

Under certain circumstances, the well-site geologist may be required to notify the intermittent services, e.g., wire-line logging, cementing, testing, etc., of when they may be needed at the well site.

## 1.3 Well-Site Geologist's Duties and Responsibilities

The well-site geologist should know his responsibilities, duties, and reporting procedures to his head office and to the well-site's engineering staff. If possible, a visit to the operating base to discuss these aspects should be arranged before the well spuds. Some operators supply special briefing packs to well-site geologists.

## 1.4 Data Acquisition

Before traveling to the rig, the well-site geologist should acquire as much relevant data as possible. This list might include the following:

—A geological and prospective background to the siting of the well
—Geological, engineering, and seismic prognoses

—Nearby well information, including mud log, final well logs, wire-line logs, and core descriptions
—Oil samples from nearby wells to compare stain and fluorescence

## 1.5 Rig Systems and Materials Familiarity

Before drilling starts or at least before the important sections are due to be penetrated, the well-site geologist should familiarize himself with the relevant rig systems and materials. This familiarization should include at least the following:

—Investigation of mud-logging equipment, particularly the following:

- Sample Catching. Be sure that a representative sample will be caught and that contamination from previous samples will be eliminated.
- Gas Analysis. Be sure that the gas analyzed is representative of the gas emanating from the borehole fluid.
- Rate of Penetration (ROP). Be sure of what and how the system is recording. For instance, on semisubmersibles and drillships the system should be tide and wave compensated.

—Investigation of rig drilling and sensing systems so that checks can be made on mud-logging sensing systems. In the event of mud-logging equipment failure, the rig drilling system sensors may be substituted.
—Investigation of mud and flow-line system
—Assembly and study of all materials that may be introduced artificially into the flow-line system

- Mud chemicals, e.g., lost-circulation materials (mica, walnut shells), lignitic materials, oil and diesels, gypsum
- Cements and additives
- Lubricants, including pipe dope and greases
- Other contaminants, e.g., pipe flake, casing shoe and plug materials

## 1.6 Logistics

The well-site geologist may be required to arrange and supervise the transport of personnel, samples, and data. Ordering procedures should be understood fully.

# 2. REPORTING AND NOTIFICATION

## 2.1 Regular Reporting

### 2.11 Daily Report

The operator may have a standard daily report form. If not, it should contain at least the following information:

—Well number and/or name
—Depth at 0600 and present operation
—Report number
—Date
—Footage drilled and a summary of operations for the last 24 hours
—Interpreted lithology with stratigraphy
—Hydrocarbon evaluation of any shows, including notification of trip, connection, and background gas
—Rates of penetration. The following guidelines should be followed:

- Hydrocarbon shows and rates of penetration should be related to their respective lithological units, which should be described separately
- The amount of detail will be dictated by operator requirements
- A copy for the on-site engineering staff may be required, and a rig copy should be retained
- The graphic lithologic log should be kept up to date and should be independent of the mud log.

See Appendix III for a sample daily report form.

## 2.2 Irregular Reporting

### 2.21 Reporting During Drilling

The well-site geologist should be sure that the drilling, mud-logging, and pressure engineering staffs are aware of their responsibilities to report drilling breaks, hydrocarbon shows, characteristic lithologies, and pressure trend deviations.

### 2.22 Final Well Report

The format of the final well and composite logs depends on operator requirements.

### 2.23 Other Reports

The well-site geologist may be required to provide other well-site reports, e.g., for cores, side-wall cores, and wire-line logging. The format and content of these reports should be discussed with operator personnel before going to the rig. See Appendices III and IV.

## 2.3 Notification

### 2.31 Services

Especially on land, the well-site geologist may be required to give advance warning to technical services such as wire-line logging.

### 2.32 Dispatch

The well-site geologist should keep a record of samples and data sent from the well site, and he should notify his base of the method of transport and the expected time of arrival.

# 3. SUPERVISION AND WITNESSING

## 3.1 Mud Logging and Associated Personnel

The well-site geologist usually is required to supervise the mud-logging personnel. He also may be required to supervise the data acquisition and pressure engineering staff. (Pressure engineering is reviewed in section 4.7.) The following points relate to mud-logging supervision:

Sample catching should be representative and uncontaminated, and the catching apparatus should be in the optimum position. The apparatus should be cleaned after every sample catching. If the shaker screens are very coarse, some provision for catching fines should be used, such as taking samples from the desander or desilter or from the flow line itself.

Samples should be lagged. Lag times should be checked repeatedly. Where any amount of uncased hole has been drilled, the theoretical lag times should be compared with actual lag times. The actual lag time may be found by introducing traceable material such as calcium carbide at connections. (Note: calcium carbide should not be used anywhere near a reservoir section.) Comparison of theoretical and actual lag times will give an indication of hole size. Early and multiple arrivals of acetylene due to carbide may indicate pipe washouts and their approximate position.

Mud loggers should be instructed to report any increase in the size and volume of cavings, any increase in gas readings, and characteristic lithologies.

Gas detection systems should be checked and calibrated regularly.

Samples should be processed (or not processed) to ensure future extraction of maximum information. Particular care should be taken to retain possible clay, fines, or hydrocarbons that might be lost while washing or drying. Samples should never be force dried.

Samples should be bagged and sealed properly and labeled with well name, number, and depth in multiple positions.

Care should be taken to ensure that the requisite number and volume of samples are being collected. Adjustments may have to be made due to drilling rates, but often extra personnel to cope with high drilling rates may be found.

The well-site geologist should check that adequate materials and spares are carried by the mud-logging unit.

The well-site geologist should ensure the quality and quantity of mud-logging recording and reporting, e.g., daily, weekly, and final well logs and reports.

The well-site geologist should supervise other duties required of mud-logging personnel, e.g., core retrieval, core packaging, and sample dispatch.

The mud-logging staff should be briefed on the confidentiality of information and should be instructed to return all such information at the end of the well.

## 3.2 Wire-Line Logging Supervision

### 3.21 General

The well-site geologist should provide details of the well to the wire-line logging crew. These will include the following:

—Well name and number
—Well location

—Elevation of drill floor
—Hole diameters and depths
—Casing information
—Mud parameters, including type, stoppage of circulation, viscosity, fluid loss, filter cake, and pH
—Hole conditions, including oversized hole, doglegs, tight spots, and deviation records

The well-site geologist should make sure that the details on the log headings are correct and that any log faults not rectified at the well site are included in the "Remarks" section.

After stopping drilling, the mud should be circulated at least twice. The last mud sample off bottom should be sampled for filtrate and mud cake evaluation. Any large deviation of mud characteristics throughout the section to be logged should be noted.

For temperature recording, two maximum recording thermometers may be run on each tool. For bottom-hole temperature evaluation, the time the last circulation was stopped and the height of the thermometer above tool zero is required.

The well-site geologist should check the depth and register of logs. The casing shoe may be used as a correlation point. Any disagreement with driller's or mud logger's TD should be reconciled immediately.

Logs should report an in-gauge hole. The overlap of logs should be sufficient, e.g., 200 feet of overlap above and below zones of interest. Master calibrations should be attached to the logs.

The correct speed and time constant should be used. A gap appears in the line at the margin of track one once per minute, so logging speed can be checked for consistency and correctness.

Maximum logging speeds are:

—BHC, 4,000 feet/hour
—Induction, 6,000 feet/hour
—FDC*/CNL*, 2,000 feet/hour

*A mark of Schlumberger.

The time constant should be recorded in the log heading.

The well-site geologist should have details of the horizontal and vertical scales to be used. Most logs are run on 1:200 and 1:500 scales. Excess backing should be avoided.

The well-site geologist should advise the logging engineer of the number of field prints required at the well site, the number of field prints and transparencies, and the edit tape format to be delivered at the base office. He should retain a reference set of field prints at the well site.

The general character of the logs should be checked.

—Logs should be one scale or a backup should appear
—Cyclic variations, zero values, and constant readings are suspicious
—Be suspicious of logs that constantly peak or level out at less than full-scale deflection
—Look especially for events that demonstrate the range of response of the tool, e.g., high- and low-porosity beds, shales, salts, anhydrite, and washouts

Acceptability of logs should be checked before breaking the tools, except where a major repair is required and a later rerun is scheduled.

## 3.22 Detailed Log Checks for Well-Site Geologists

The checks included are designed to sort out many of the problems encountered in wire-line loggings. However, they are not as comprehensive as those which a log analyst at the well site might investigate.

### Gamma Ray

**PRINCIPLE**
—Measurement of formation natural radioactivity

**USES**
—Definition of bed boundaries
—Assistance in lithological identification

—Calculation of degree of shaliness
—Well correlation

**CHECKS**

—Near-zero API unit readings are uncommon except in massive halite, gypsum, or anhydrite
—Most coals are less than 50 API units
—Some radioactive shales and potash generate more than 500 API units
—Uraniferous beds may generate readings outside the range of the tool
—API values may vary due to tool running speed and hole and mud variations
—Static drift should be checked before and after calibration counts
—Repeat sections may not agree in absolute values, but the curve character should be similar

## Spontaneous Potential

**PRINCIPLE**

—Measurement of formation potential

**USES**

—Identification of bed boundaries
—Identification of permeable zones
—Calculation of formation salinity from $R_w$
—Estimation of shaliness

**CHECKS**

—Drift to the left due to temperature change is normal
—If a base-line shift is required, it should be done in a shale section
—The SP log should be compared with nearby wells
—The SP log should be a smooth curve with no rapid or cyclic variations

## Caliper (Mechanical)

**PRINCIPLE**

—Measurement of size or shape of hole with mechanical arms

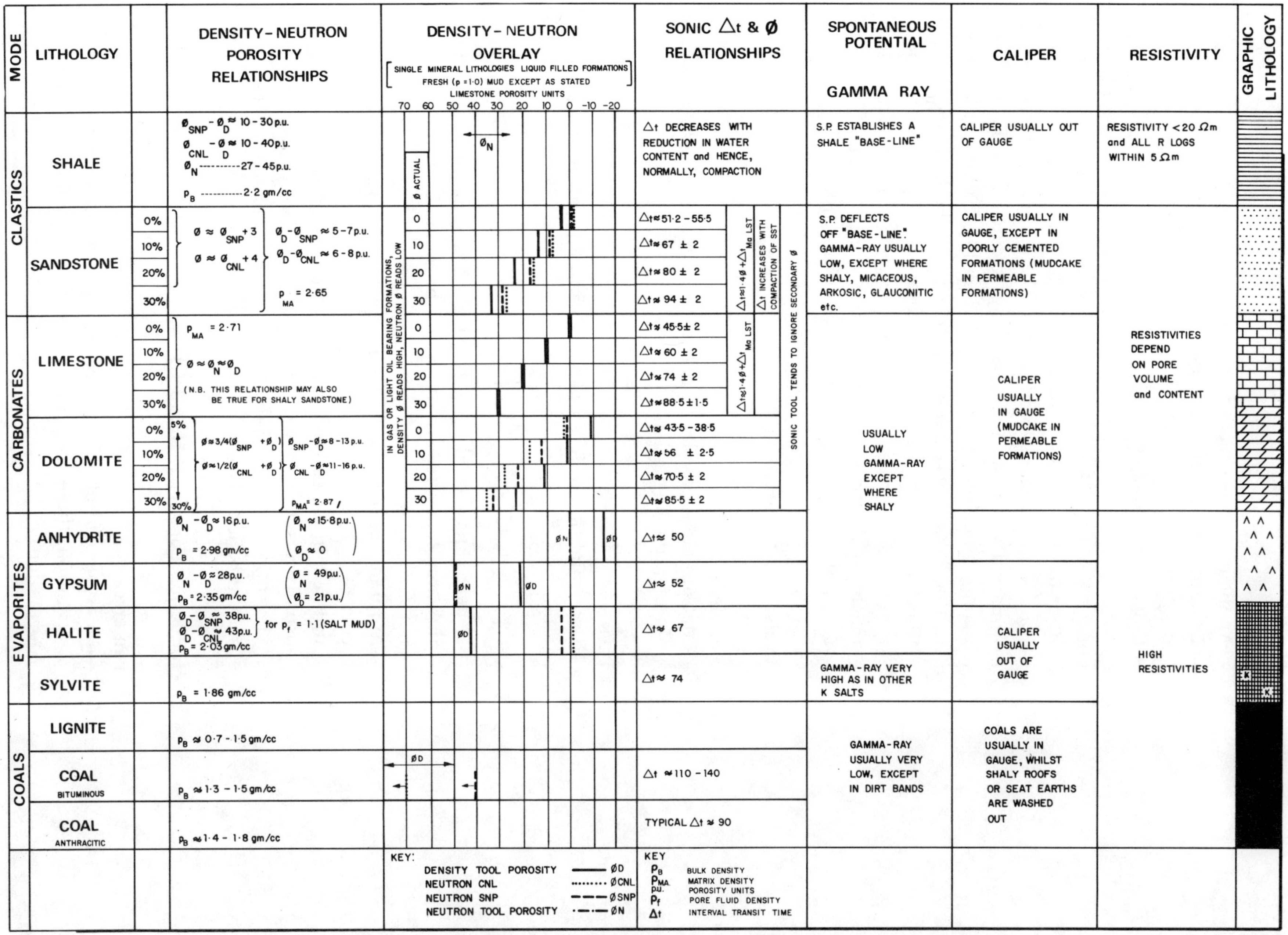


**Fig. 1** Well-site lithology from logs, derived from Schlumberger data

**USES**

—Application of corrections to logs
—Location of good hole conditions for geophone, packer, etc., position
—Calculation of cement volumes

**CHECKS**

—Repeatability is not a relevant check
—Hard, dense zones should be near the nominal bit size
—Apparent washouts giving readings greater than the maximum caliper extension are usually erroneous, except in salt beds, and may point to an electronic fault
—Check for general agreement with other caliper logs but be aware of hole irregularities, the presence of mud cake, and mechanical attributes of different calipers
—Check the caliper log in the casing if possible

## Resistivity Electric Log (ES)

**PRINCIPLE**

—Measurement of voltage, hence resistivity, by passing current through the formation

**USES**

—Determination of $R_{xo}$ and $R_t$. This tool is rarely run any more.

**CHECK**

—Repeat runs should agree to within 5% of the full-scale value
—In water zones, resistivity should be a small but finite value, not zero
—Amplified curves should be in register with the regular curves
—In thick, in-gauge shale sections, the short normal curve usually has a slightly higher value than the other curves. The ratio of values of the long and short normal curves should be less than 2:1 in such sections.

## Induction Logs ($IL_d$, $IL_m$, LL8)

**PRINCIPLE**

—Measurement of induced secondary currents in the formation; these are proportional to the conductivity of the formation

**USES**

—Determination of $R_{xo}$ and $R_t$

**CHECKS**

—Repeatability in zones of interest is very important
—In thick compacted shales, ($R_{IL}$ medium/$R_{IL}$ deep) $< 2$
—There should be no zero readings

## Laterolog* Logs (Excluding Microlaterolog* Logs)(LL9)

**PRINCIPLE**

—Measurement of voltage, hence resistivity, by passing current through the formation

**USES**

—Determination of $R_{xo}$ and $R_t$

**CHECKS**

—Resistivity in pipe should be near zero
—In thick, compacted uninvaded shales, ($R_{IL}$ deep/$R_{IL}$ shallow) $< 2$

## Contact Resistivity Logs (Microlaterolog*, MSFL*)

**PRINCIPLE**

—Measurement of voltage, hence resistivity, by passing current through the formation

**USES**

—Determination of $R_{xo}$
—Delineation of permeable beds by detection of presence of mud cake

**CHECKS**

—There should be no readings near zero except in washouts with salt muds
—Resistivity values should be greater than mud resistivity

---

*A mark of Schlumberger.

—Caliper should be checked in the pipe
—A mud log may be recorded running into the hole with the tool closed; this will verify $R_{mf}$ and tool operation

**Compensated Acoustic Logs (BHC* Sonic)**

**PRINCIPLE**

—Measurement of acoustic travel time

**USES**

—Lithological determination, especially in evaporites
—Porosity evaluation

**CHECKS**

—Look for cycle skipping and spikes
—Values of less than 40 μsec/foot and greater than 190 μsec/foot are unusual
—Check the register of the sonic log with the gamma ray
—In clean, porous water zones without vugular or fracture porosity, $\phi$ sonic should agree with other porosity values to +/− five units
—Ensure that integrated travel time is recorded as pips in the margin of track two
— Sonic reading should be 57 μsec/foot in the casing

**Compensated Density Logs (FDC*)**

**PRINCIPLE**

—Measurement of energy loss between a radioactive-emitting source and return of gamma rays to a detector. The loss of energy is by collision of gamma rays with electrons in the formation, the electron density being directly proportional to the bulk density.

**USES**

—Porosity evaluation
—Measurement of formation density, especially in evaporites
—Gas detection
—Evaluation of hydrocarbon density
—In combination with a neutron log, it may be used to identify lithology and to estimate shale content

---

*A mark of Schlumberger.

**CHECKS**

—In clean, porous water zones, $\phi$ density should be compared with other porosity values
—In pure anhydrite, gypsum, salt, or dense carbonate beds greater than 36 inches thick, the zero porosity value should be checked
—The $\Delta\rho$ should be mainly positive except in dense baritic muds or gas zones. Consistent offset or drift above zero is suspicious
—Repeatability should be to within 0.05 grams/cc except in washed-out zones

## Side-Wall and Compensated Neutron Logs (SNP, CNL*)

**PRINCIPLE**

—A neutron source bombards the formation and the resultant scattering and neutron deceleration depend largely on formation hydrogen content. Epithermal neutrons are detected.

**USES**

—Porosity evaluation
—Delineation of porous formations
—Evaluation of shale content
—In combination with FDC* logs, they may be used to identify lithology and to estimate shale content

**CHECKS**

—Leapfrogging is caused by tool sticking and is indicated by a sawtooth appearance of the caliper over a short distance. If in doubt, a tension curve should be requested
—In gas zones, $\phi$ neutron $<$ $\phi$ density
—In tight and clean limestone, anhydrite, or dolomite, neutron should be checked for zero porosity
—$\phi$ neutron should be checked with other porosity measurements

## Dipmeter (HDT*)

**PRINCIPLE**

—Comparison of three or four identical microresistive devices whose results are analyses to obtain formation dip

**USES**

—Determination of sedimentary and/or structural dip

**CHECKS**

—Individual traces should be identical and slightly out of register

—Good detail should be shown on the curves

—Rate of rotation of the tool, as detected by the azimuth of electrode 1, should be less than one turn/minute/50 feet

### Cement Bond Log (CBL)

**PRINCIPLE**

—Measurement of acoustic velocity

**USES**

—To check cement bond integrity

**CHECKS**

—Amplitude and total travel time curves should correlate

—A straight-line amplitude from cement top to TD indicates a poor sensitivity selection

—Repeatability should be to the +/− chart division; deviation from this may be due to poor tool centralization

—Check the free pipe signal by running above the known top of the cement

## 3.3 Side-Wall Core Retrieval Supervision

During shooting, depths should be checked by the well-site geologist, who should be present in the logging unit

The well-site geologist should supervise the correct placement of samples in the bottles. Bottles should be marked with the well name and number and the depth. The depth should also be marked on the cap.

A note of recovery should be kept. The well-site geologist may be required to agree with the logging engineer as to which samples are to be paid for. (Good recovery is about one inch in soft formations and slightly less in hard formations.)

The well-site geologist may have to decide on side-wall core reruns. This should only be considered over important sections of the hole and where the caliper log suggests a good chance or recovery.

The drilling supervisor should be advised of any bullets left in the hole since they are potentially obstructive.

## 3.4 Velocity-Shot Supervision

### 3.41 General

The procedure for picking velocity shot points is given in section 4.6.

The contractor should have a preshoot check procedure to determine if his apparatus is in working order. At the least, this should check the following:

—The downhole tool-clamping mechanism is operative
—Gun firing and pressure gauges are operative to standard
—The depth recorder should be checked for zero by suspending a well geophone at the rotary table
—The gun hydrophone should be checked for a clear time break register for time-zero pinpointing
—A general idea of velocity shot geometry is given (see Fig. 2)

### 3.42 Noise Sources

The following noise sources should be considered:

—All unnecessary rig machinery should be shut down
—As far as possible, all extraneous noise sources should be eliminated
—Offshore shooting should be done when sea-swell noise is at its lowest. Lulls of a few seconds' duration or more may occur every 30–60 seconds. If they do, shoot during these lulls. If they do not, lower the geophone a few feet and reclamp.

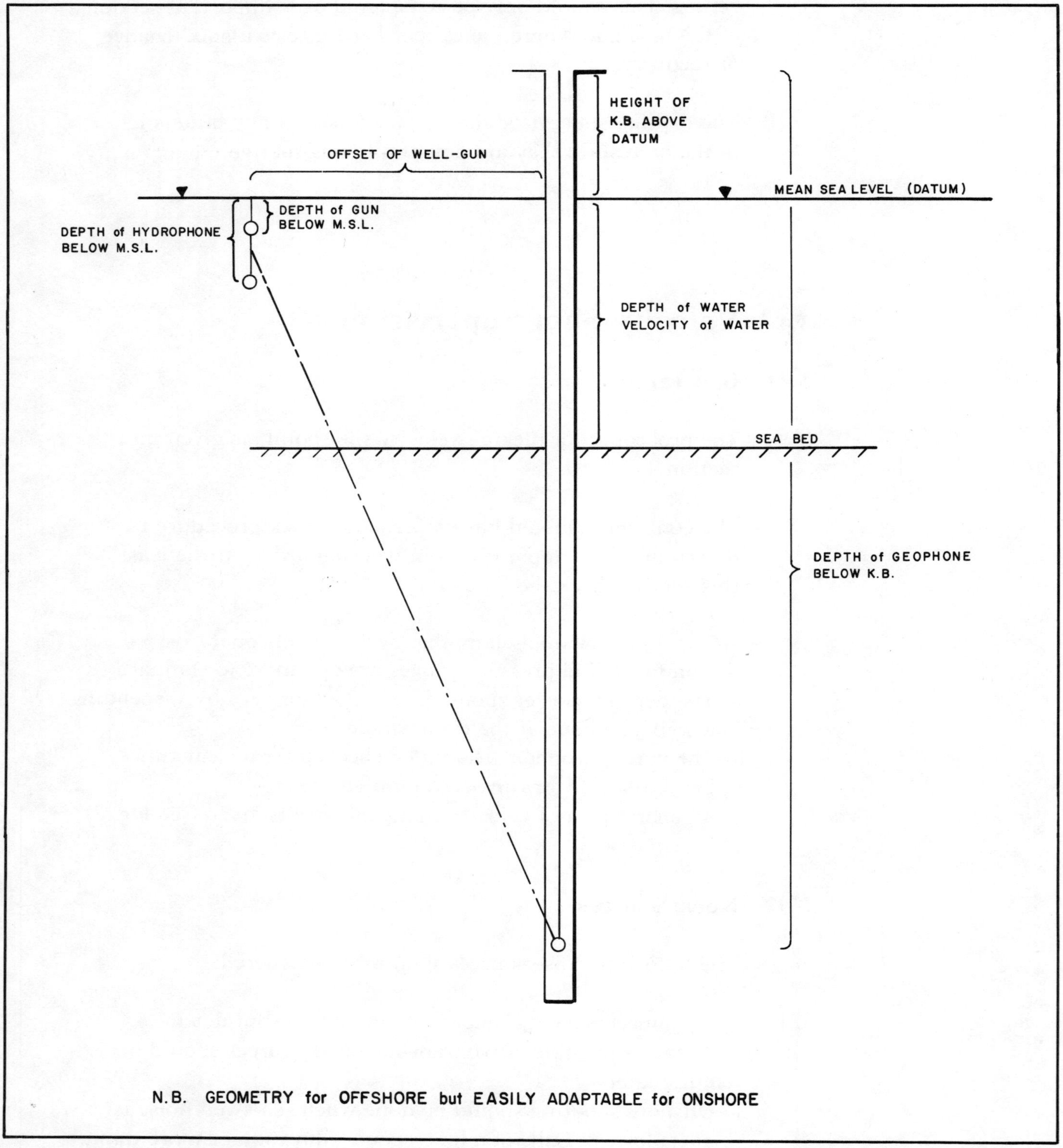


**Fig. 2** Velocity shot geometry

### 3.43 Depth Recording

This should be monitored continuously, and at least the following checks should be carried out:

—Well geophone should be zeroed at the rotary table
—Winch readings should be checked against the magnetically activated bell
—At the end of the survey, the zero should again be checked by suspending the well geophone at the rotary table

### 3.44 Shooting

The following points should be noted when shooting:

—Shooting should be initiated from the top of the hole
—Where the survey is in two parts, there should be an overlap (say three shots) of repeat shots

### 3.45 Shot Recording

It is recommended that three good shots be taken at each level. The criteria for a good shot are as follows:

—Clear arrival time
—Repeatability of results to 1/2 ms. Check tool clamping if this is not forthcoming

### 3.46 Information Recording

Be sure to note the following points when recording information:

—Retain one analog record from each level
—Note checklist geometry before the end of the survey
—Offshore sea-level variations at and during the survey should be noted
—Compare calculated interval velocities with the sonic log to check the correct timing

## 3.5 Core Retrieval Supervision

The operator's policy on core retrieval should be followed. In the absence of any policy, the following procedure is recommended.

### 3.51 Preparation

—Check that all requisite materials are present
—Ensure that sufficient core boxes are numbered

- Due to core breakage, a greater length of core boxes than the maximum nominal length of the core may be required
- To facilitate marking and subsequent identification of core labeling, white paint may be used as a base for marking core boxes
- All boxes should be labeled with T and B to identify the top and bottom of the core section
- All boxes should be labeled with the well name and number
- All boxes should be labeled with the core number and the box numbers

—Plastic bags should be at hand for broken cores
—The use of intermediary trays for retrieval from the barrel and before packing into core boxes is not recommended because it adds another stage in the process and increases the chance of error. However, if they are included, these too should be labeled clearly and correctly

### 3.52 Retrieval

The well-site geologist should be clear about how the retrieval is to proceed. Two people may be used: one to retrieve and pack the core and one to remove and replace core boxes. It is recommended that the basal part of the core goes into the bottom of box 1 and that the top of the boxes are nearest the core barrel itself (see Fig. 3).

Rig time is expensive, and so is coring. It is worthwhile spending extra minutes to ensure correct assignation of depths to the core.

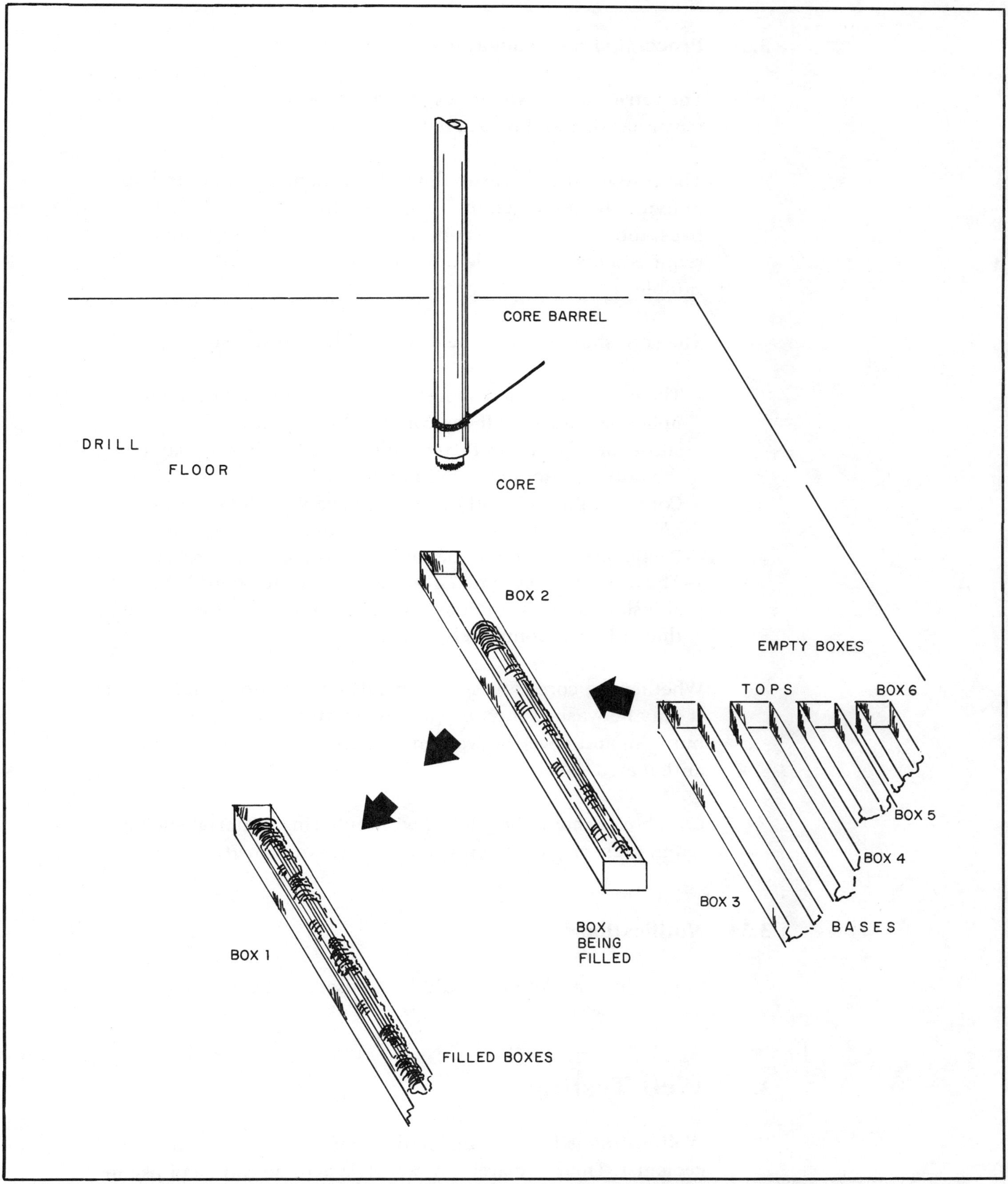


**Fig. 3** Recommended core retrieval system

### 3.53 Processing and Packaging

The retrieval core should be taken somewhere quiet and secure for processing and packaging.

The core must be measured, marked, examined, and perhaps individually packaged in aluminum foil, cling film, plastic bags, tubing, wax, or seal-peel according to the operator's requirements. "This side up" should always be marked on the sample.

The core should always be examined at the well site.

—The core description should be carried out as described under section 4.12. If the core has been disturbed subsequently, the well-site geologist's description may be the only key in unraveling the problem.
—Core examination and description should be carried out by both the well-site geologist and the mud-logging staff
—Particular note should be taken of hydrocarbon indications
—The amount of core taken for destructive testing at the well site should be as small as possible. Some operators require that it be left completely intact.

Whether the core is washed depends on operator requirements. It may be washed if required only for stratigraphic purposes, but it should never be washed if hydrocarbon evaluation is anticipated.

Core boxes should be packed with buffering material such as foam, rags, or paper to prevent damage in transit.

### 3.54 Notification

Core shipment details should be advised to shore base.

## 3.6 Well Testing

Well testing is largely outside the sphere of the well-site geologist. There are areas, however, where he can be of use in the testing operation.

In uncased holes, the tools run include the wire-line FLT, RFT, and occasionally the DST tools. In cased holes, the DST and its variants are used to test the well. Offshore, DSTs are almost always run only in cased holes.

The FIT, the formation interval tester, may be used to test the contents of permeable zones (although the zone may be too invaded by filtrate to give meaningful results) and to obtain formation pressures. The RFT, the repeat formation tester, is similar to the FIT but may take multiple pressure readings and two or three samples. Therefore, the RFT may be used better to pinpoint hydrocarbon contacts by monitoring pressures and collecting samples.

Because the FIT and RFT tools are run in an open hole and are wire-line logging tools, the operation is relatively cheap and may be run before a decision to run casing is made. The well-site geologist may be involved in these tests through his knowledge of hydrocarbon shows, lithology, and quick-look analysis of electric logs.

The DST, the drill-stem test, is a more involved operation and is usually run only in cased holes. The well-site geologist normally will not remain at the well site for the duration of such a test. However, he can help by having at his disposal all relevant information such as the location of hydrocarbon shows, lithology, and electric log for liaison between the rig and the head office.

## 3.7 Pressure Engineering Supervision

### 3.71 Background

#### Definition

The term overpressure is commonly used, but this is very often misleading. A better term is abnormal pressure, since we are concerned with differential pressures and pressure gradients.

#### Abnormal Pressure Situations

Areas of formation pressures greater than hydrostatic tend to be concentrated in young geosynclinal areas where shale deposition has been rapid and sandstone deposition has been very minor. Reservoir geometry may be conducive to the generation of higher-than-normal pressure, e.g., there may be pressure transmitted from water at the base of the reservoir to oil and particularly gas above it. There may be a pressure communication from a level below that being drilled, produced by faulting or artificially by an improperly plugged well.

Under artesian conditions, higher-than-normal pressures may be encountered. Conversely, where drilling into a formation with a surface connection below that of the drilling location, subnormal pressures may be encountered.

Secondary recovery methods in nearby wells may generate artificially elevated pressures.

Subnormal pressures may occur where a reservoir has been depleted or by the sealing and burial of a fully dewatered formation. Very much higher-than-normal pressures may be generated by physical chemistry changes in evaporites.

#### Importance of Recognition

Under conditions of imbalance between a reservoir and the borehole fluid, the reservoir may be damaged. Conversely, reservoir fluids may flow into the borehole. Gas or water emanating from a thin sandstone in an overpressured shale section may initiate a disastrous train of events. Where the borehole pressure is greater than that of the formation, formation damage and lost circulation may occur. Where formation pressure is greater than that in the borehole, there will be a tendency for sloughing, which may hamper lithological identification. The chances of drilling a safe and informative well are thus very much enhanced by knowledge of pressure regimes.

### 3.72 Pressure and Monitoring Evaluation

The following physical-chemical attributes may be affected by a varying pressure regime.

—Compaction
—Porosity and water content

—Pressure-dependent fluid movement parameters such as ionic concentrations and hydrocarbon saturations
—Differential pressure between the borehole and the formation

The aim of pressure engineering is to measure the parameters that reflect these changes. They may be grouped under evaluation and prediction. Evaluation should be carried out at the prewell planning stage and should be reevaluated at the end of the well. Prediction, which can be done at the well site, aims to forecast abnormal trends at or around the time they are drilled.

### Single-Value Direct Readings

Formation tests such as DST, FITs, and RFTs can give good values for reservoir pressures but are rarely used in overpressured zones due to operational risks under poor hole conditions.

Kicks can be used to calculate formation pressures.

### Continuous Predrilling Readings

#### SEISMIC VELOCITY

In a thick, constant shale section, seismic velocity will depend largely on water content. In theory, an overpressured zone is indicated by a decrease in seismic velocity. Interpretation by this method is difficult and of little use in wildcat areas (see Fig. 4).

### Continuous Postdrilling Readings

#### WIRE-LINE LOGS

Resistivity and conductivity logs may be used to detect the higher water content in overpressured shales. The deeper investigation tools will give better results.

Pososity logs may also be useful in detecting overpressured shales. The sonic tool is usually the best in this respect. In all cases when using logs to evaluate pressure regimes, care should be taken to ensure consistent lithology. Hence, gamma-ray and shot-point logs may be used advantageously (see Fig. 4).

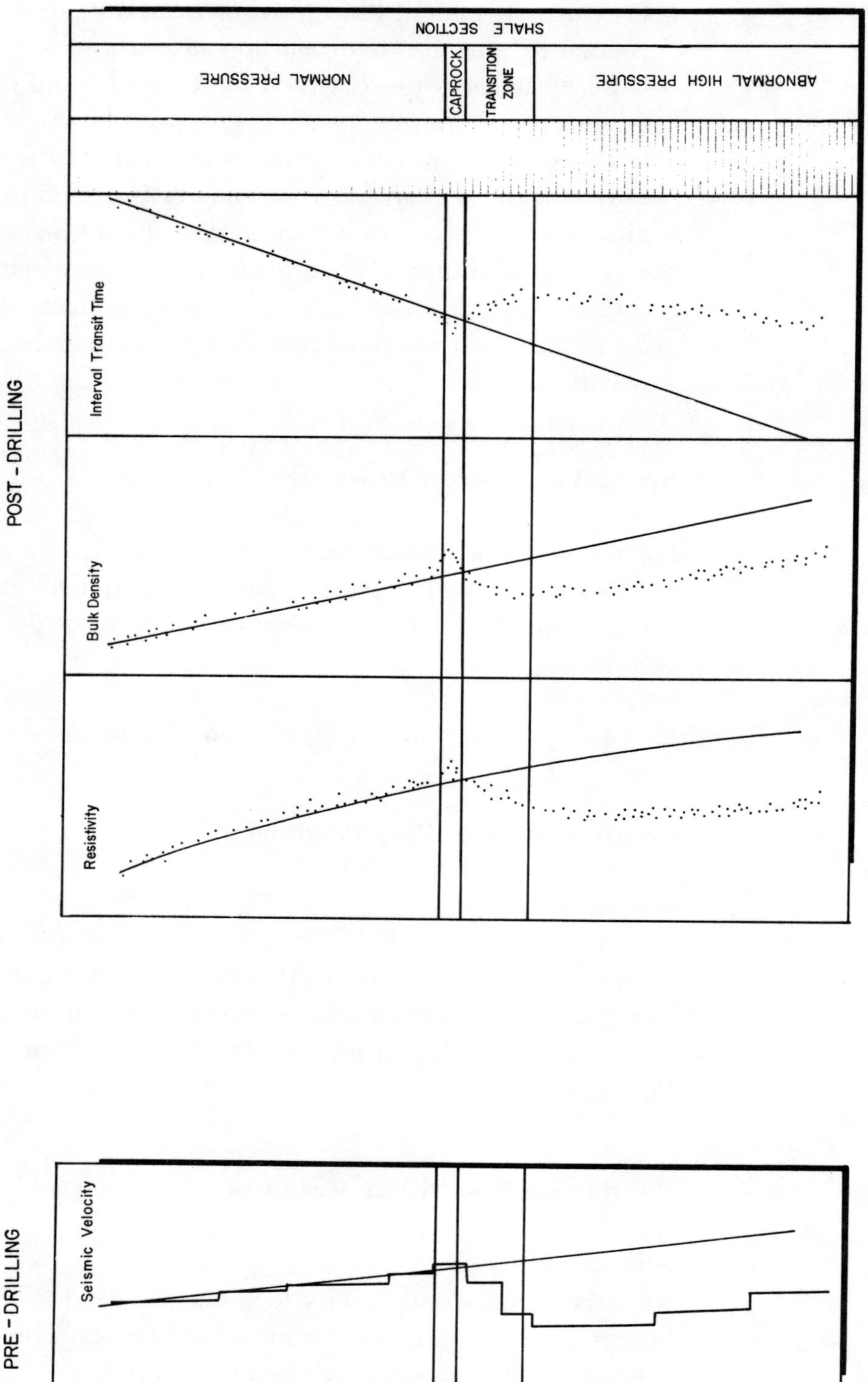


**Fig. 4** Pressure parameters, pre- and postdrilling

## Continuous During-Drilling Readings

#### FLOW-LINE TEMPERATURE

The aim of this method is to search for a decrease in thermal gradient in the transition zone and an increase in gradient in the overpressured zone itself. This arrangement of temperature gradient is caused by the increased water content in the overpressured zone where the thermal conductivity will be reduced. There are problems associated with this method. Permeable beds interbedded in the shale cause temperature anomalies. Offshore, the marine riser will cause attenuation of temperature differentials. Interruptions in circulation caused by trips and connections will require the temperature plot to be smoothed (see Fig. 5).

#### PENETRATION DATA

Normalized penetration rate formulations take care of any change of penetration rate due to a change in drilling parameters. Allowances may also be made for hydraulics.

Before concluding that a penetration rate trend is changing, one must carry out the foregoing corrections, check that the lithology is constant, and finally estimate if the deviation may be due to bit wear. Only then can one reasonably assign a change in penetration to a change in pressure regime (see Fig. 5).

#### BULK DENSITY

In a constant thick shale section, a plot of bulk density may be used to observe change in pressure regimes due to change in compaction and the difference in water content. Although the lithology appears to be constant, some shales may react differently to the mud fluid, giving anomalous readings (see Fig. 5).

#### GEOCHEMICAL METHODS

Many geochemical changes occur in switching from one pressure regime to another. The combined effect of these changes may be measured by a differential conductivity probe.

Individual methods of testing shale cuttings include the following:

—Shale factor measures the cation exchange capacity and hence its diagenetic state

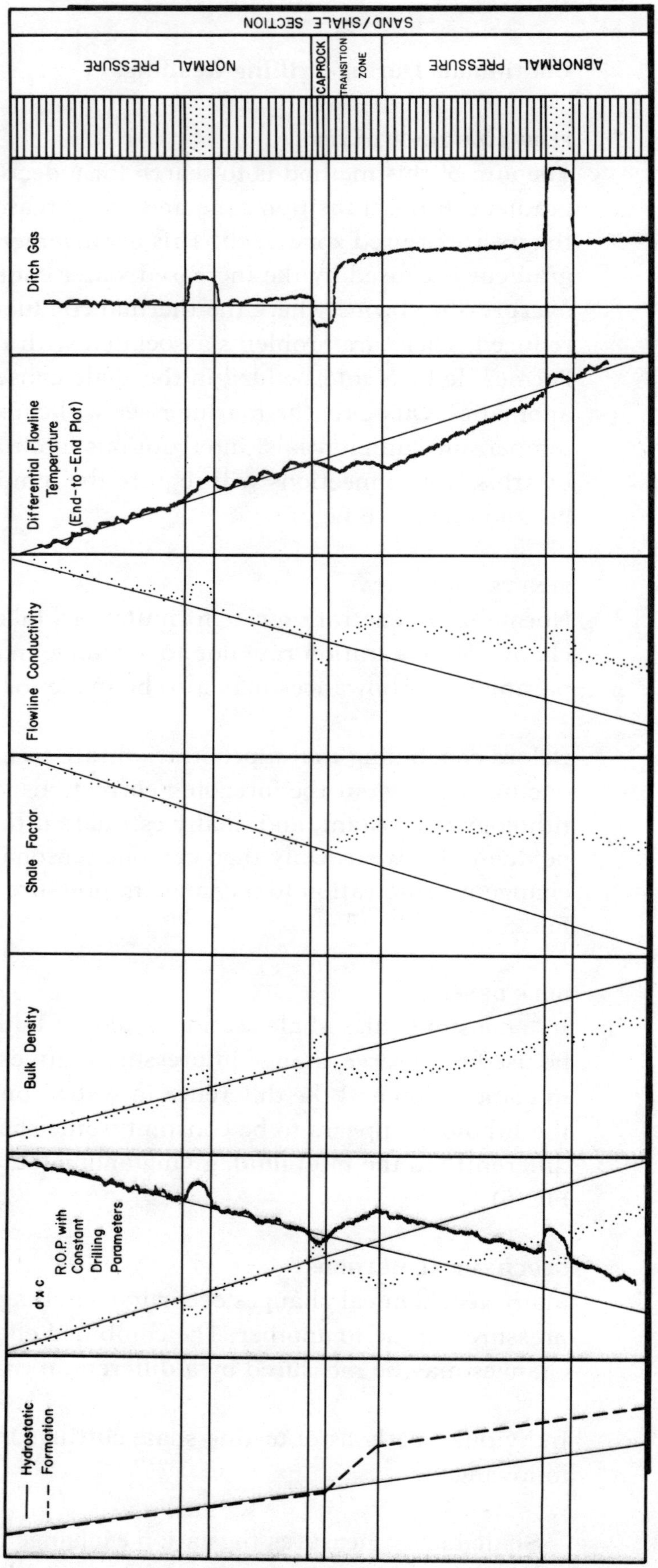


**Fig. 5** Pressure parameters during drilling

—The concentration of bicarbonate ions is relatively high in a cap rock and in an overpressure zone is still high
—The redox potential above an overpressured zone will change the trend to an increase that will be maintained in an overpressured zone. Where permeable formations occur, there will be complications in evaluation
—Shale cuttings may be tested for water content by various methods; however, varying reactions with mud fluid may complicate the evaluation

**BOREHOLE CONDITION**

The following indications show that the borehole pressure is insufficient to contain formation pressures:

—Increasing trip and connection gas
—Increasing drag on pulling up
—Swabbing on pulling out of the hole
—Increase in size and volume of cavings
—Increase in torque when drilling
—Occurrence of bottom fill
—Lingering gas from thin permeable zones
—Increasing background gas

## 3.73 Reporting and Liaison

It normally is not a geologist's job to construct a pressure log. However, his evaluation of the drilled lithology is vital to any meaningful pressure log. The geologist's decision may influence whether a deviation from a trend will be acted upon, so he should be aware of these implications of any pressure data available.

Liaison with drilling, mud, mud-logging, and pressure engineering staff is important, and it is vital to assemble all available data.

Due to the possibly severe effects of misinterpretation of an overpressure zone, the maxim should be: "Report trend deviations first; answer questions later."

## 3.74 Summary

The value of a decision on pressure regimes is enhanced by the geologist's knowledge of pressure engineering, his assembly and consideration of all relevant data, and an accurate interpretation of lithology. (A sample blank pressure evaluation report form is included in Appendix III.)

# 4. WELL-SITE DUTIES

## 4.1 Lithological Evaluation

### 4.11 General

Lithological evaluation and recording should be carried out in an orderly and objective fashion (see Fig. 6). The depth of lithological evaluation obviously will depend on the situation, available time, and operator requirements, but the following procedure designed for analysis of cuttings, cores, or side-wall cores is aimed at producing a fairly comprehensive practical study.

—Make clear in the recording of the lithology just what is being evaluated:

- A single sample
- A spread of samples, e.g., a tray of cuttings
- An interpreted lithological type
- A spread of interpreted lithologies, e.g., for a 24-hour period

—Ensure that the order of the description is constant
—Observe one unwashed sample with one washed sample so that clay and silt content may be analyzed accurately
—Ensure that sample washing is carried out in a consistent manner
—Never force dry a sample
—Keep a running show of previous samples so that subtle lithological changes can be noticed
—If the volume of cavings is high, corrective measures should be requested and noted when they are carried out

1. ROCK TYPE
2. COLOUR
3. TEXTURE
    a) Grain size
    b) Physical texture
    c) Sorting
    d) Roundness
    e) Sphericity
    f) Carbonate classification
4. HARDNESS and FRACTURE
5. CEMENTATION
6. ACCESSORIES
7. POROPERM
8. HYDROCARBON INDICATIONS
9. OTHER FEATURES
10. ARTIFICIALLY INTRODUCED MATERIALS

**Fig. 6** Lithological order of description

## 4.12 Evaluation and Description

### Lithology

The rock type lithology may be qualified adjectively if individual subordinate constituents comprise more than 20%, e.g., sandy limestone. Where the proportion is less than 20%, they are classed as accessories. The following tests should be carried out to identify mineralogy and hence rock type:

#### HCl TEST FOR CARBONATES

The rate of attack, the uniformity of attack, the strength of the acid, the temperature of the acid, residue, and any other peculiarities should be noted (see Fig. 7). Tests may be carried out on a spot plate or test tube.

—The stronger the acid, the quicker the reaction will be. Dolomite may require strong acid for any reasonable reaction
—The action of heat will speed the reaction. The presence of a light bulb close to a spot plate may produce a significant amount of heat
—Inhomogeneity, as with dolomite rhombs in limestone, may produce variable reactions; therefore, the reaction should be observed microscopically
—The action of crushing will increase the surface area and hence increase the rate of reaction
—Note the oil reaction. Hydrocarbons (either natural or artificial) may form a film over carbon dioxide bubbles, increasing their size and making the sample float. This test is very sensitive.

#### CARBONATE STAIN TEST

This test is used to differentiate between various carbonate minerals (see Fig. 8a). The reagent is:

—0.2 percent HCl
—0.2 percent alizarin red
—0.5 percent saturated potassium ferricyanide

#### PHYSICAL ATTRIBUTES

—Density. For example, bromoform may be used to differentiate between gypsum and anhydrite. Bromoform

| LITHOLOGY | % $CaCO_3$ | REACTION |
|---|---|---|
| Limestone | 90-100 | i) Instant Violet reaction |
| | | ii) Specimen is buoyed by frothing carbon dioxide gas, and floats on the surface |
| | | iii) Dissolution in five minutes |
| Dolomitic Limestone | 50-90 | i) Moderate but continuous reaction |
| Calcareous Dolomite | 10-50 | i) Weak, but increasing reaction |
| Dolomite | 10 | i) Hesitating reaction up to thirty minutes |
| | | ii) Acid turns milky |

*N.B. Note the oil reaction. Hydrocarbons will form a film over $CO_2$ bubbles, increasing their size and producing flotation of sample.*

**Fig. 7** 10% HCl reaction

| | | |
|---|---|---|
| RED | Pure Calcite | |
| MAUVE | Ferroan Calcite-----$Fe^{2+}$ | Poor |
| PURPLE | Ferroan Calcite-----$Fe^{2+}$ | Rich |
| LIGHT BLUE | Ferroan Dolomite | |
| DARK BLUE | Ankerite | |
| UNSTAINED | Pure Dolomite | |
| REAGENT | 0·2% HCL | |
| | 0·2% ALIZARIN RED | |
| | 0·5% SAT. K FERRICYANIDE | |

| | | | |
|---|---|---|---|
| Lime Mudstone | Mud Supported | Grains | 10% |
| Wackestone | Mud Supported | Grains | 10% |
| Packstone | Grain Supported | Mud | 5% |
| Grainstone | Grain Supported | No Mud | |
| Boundstone | Components bound together during deposition | | |

**Fig. 8** Carbonate stain test and Dunham carbonate classification

has an SG of 2.5; therefore, gypsum floats and anhydrite sinks in it.

—Hardness
—Fracture
—Color
—Color of streak
—Texture, including crystal shape

## Color

The color description should conform to the company's standard legend. A complete color definition may be given by comparison with Munsell color charts, taking care to view the sample under the correct light criteria.

The sample should be observed wet to increase the intensity of its color. Wetting the sample will not, however, change its color saturation.

Where the color is not uniform, a note should be made of this.

The color texture of the sample should also be described, e.g., it may look earthy or soapy.

## Texture

### GRAIN SIZE

Use a microscope with a grain-size comparator or graticule (see Fig. 9a).

### PHYSICAL TEXTURE

The physical texture should be noted, e.g., vitreous or earthy. The swelling properties of clays and shales should be noted.

### SORTING

For a definitive laboratory evaluation of grain sorting, sieve analysis should be carried out. At the well site, however, use is usually made of visual comparators or tables. Note should be taken of the sorting mode, e.g., bimodal.

### ROUNDNESS

Use the chart in Fig. 10a.

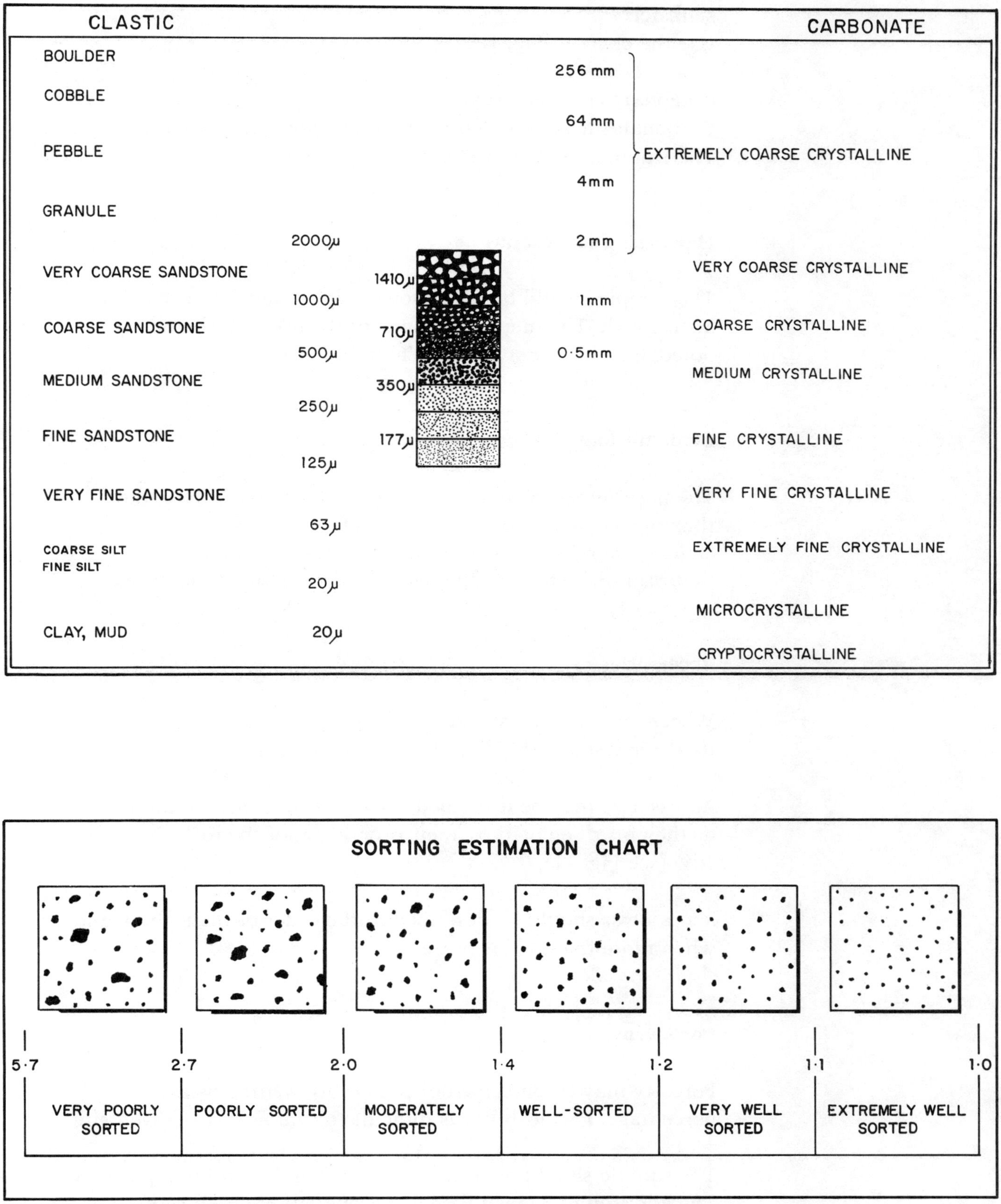


**Fig. 9** Grain size estimation chart and sorting estimation chart

#### SPHERICITY

Use the chart in Fig. 10b.

#### CARBONATE CLASSIFICATION

Carbonates may be classified by their mode of grain support. The Dunham system is shown in Fig. 8b.

### Hardness and Fracture

The sample should be tested for hardness and fracture using a steel pick. The mode and shape of the fracture should be noted, e.g., brittle or conchoidal (see Fig. 10c).

### Cementation

The pore cement material should be described. Where more than one cement is present, the relative ages of individual cements should be noted. Where pressure solution has occurred, the effects of the degradation should be described.

### Accessories

Where minerals individually comprise less than 20% of the total, they should be described in this section.

Accessories may be described, such as trace, rare, or abundant, or they may be noted by their proportion of the total, e.g., 10% (see Fig. 11).

Accessories should be described in the order of their abundance when more than one is present.

### Poroperm

Porosity may be described or preferably written as a percentage. Permeability is difficult to evaluate at the well site.

The sample should be observed dry when determining porosity. The addition of water to a dry, porous sample may be used as an indicator of porosity.

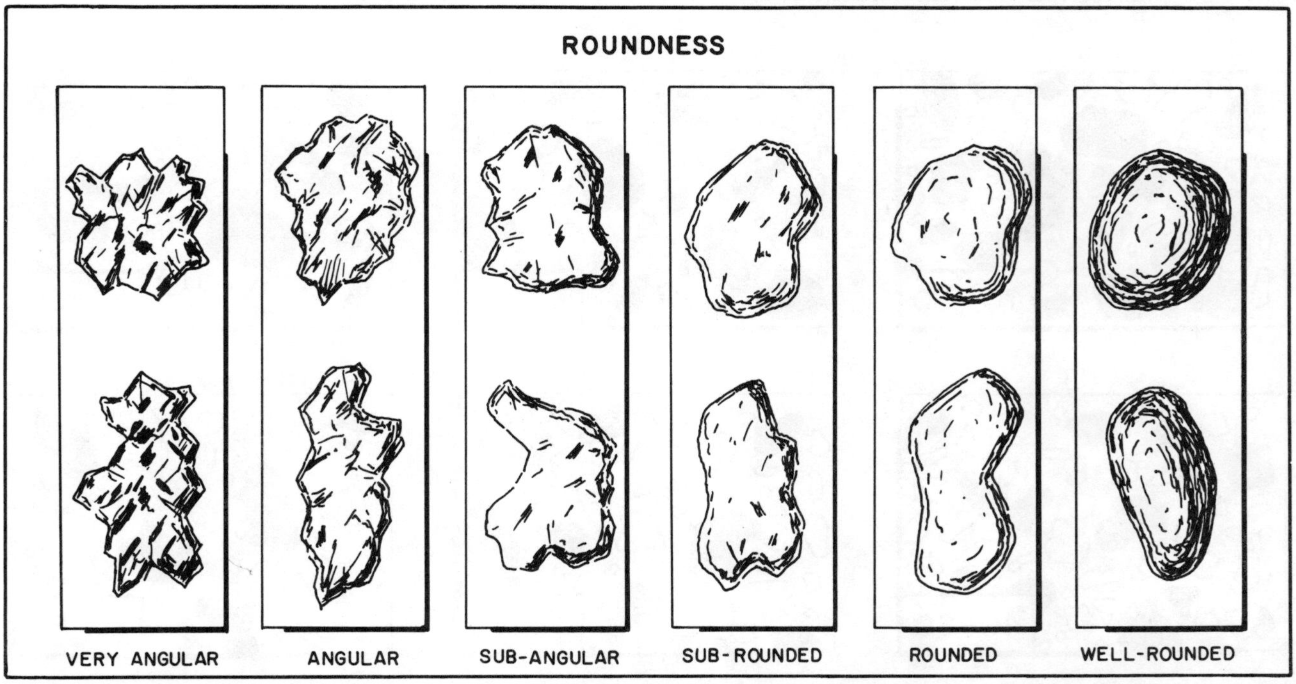


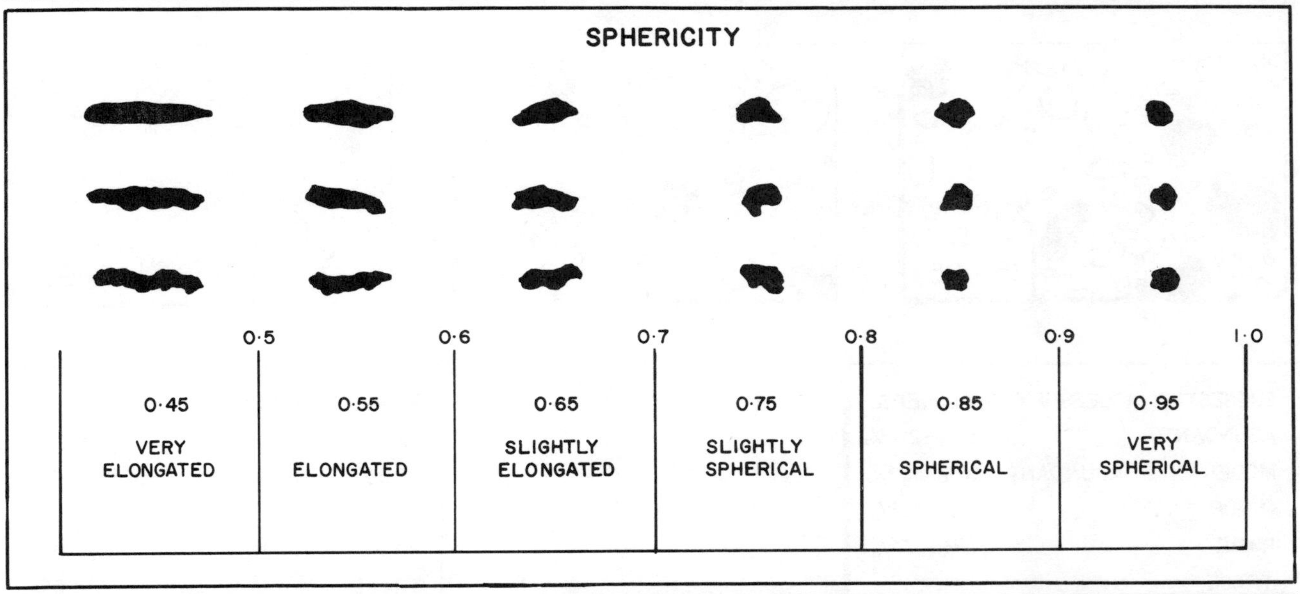


**HARDNESS**

| MOH'S SCALE HARDNESS: | STANDARD MINERAL: |
|---|---|
| 1 | TALC |
| 2 | ROCK SALT, or GYPSUM |
| 3 | CALCITE |
| 4 | FLUOR-SPAR |
| 5 | APATITE |
| 6 | ORTHOCLASE FELDSPAR |
| 7 | QUARTZ |
| 8 | TOPAZ |
| 9 | CORRUNDUM |
| 10 | DIAMOND |

SUGGESTED HARDNESS QUALIFIERS:

LOOSE: Grains fall apart in dry condition
FRIABLE: Rock crumbles easily between fingers
MODERATELY HARD: Grains can be detached with pick
HARD: Grains cannot be detached with pick
SOFT:
PLASTIC: e.g. clay
BRITTLE: e.g. coal

**Fig. 10** Roundness, sphericity, and hardness charts

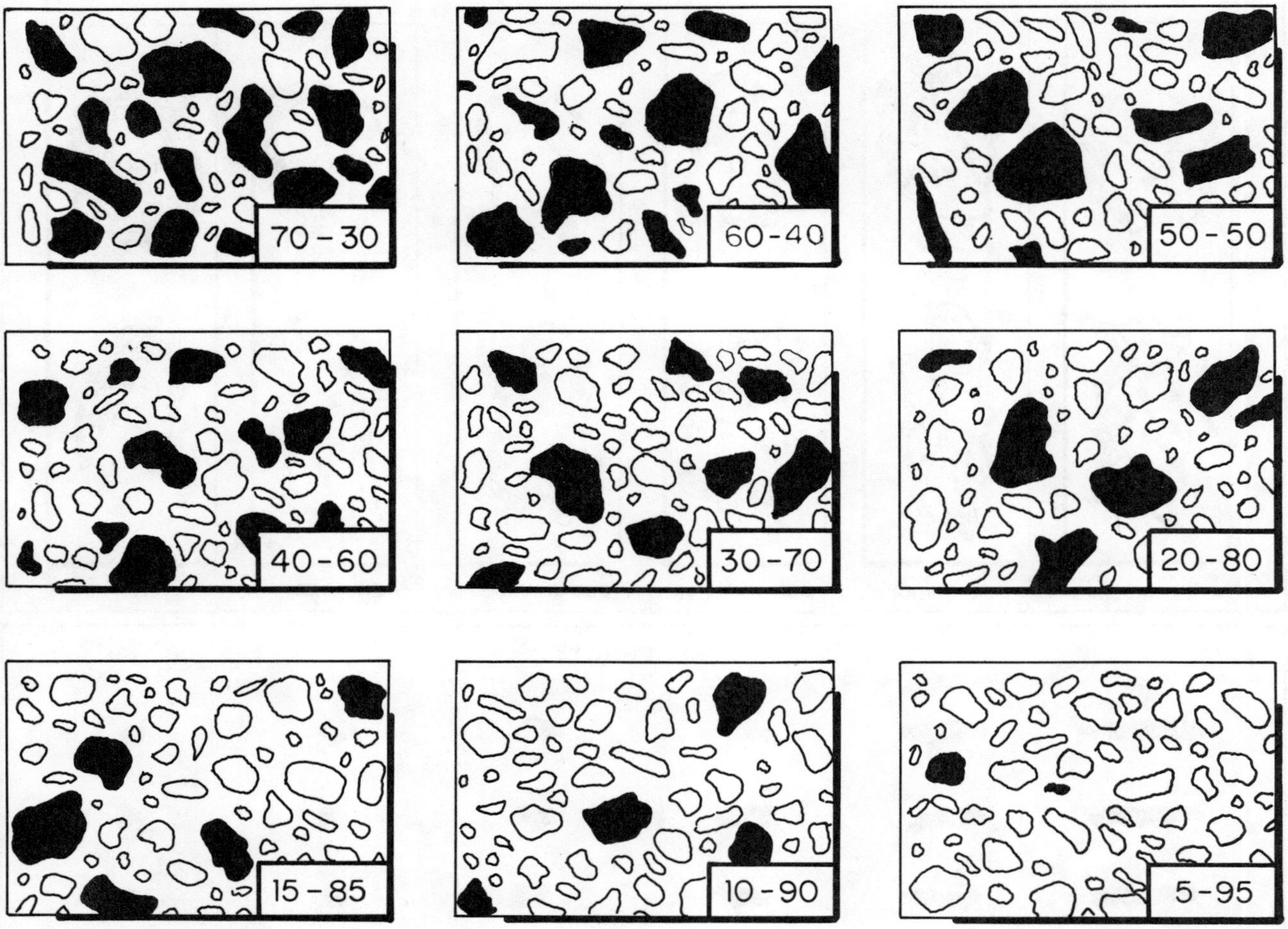


| SUGGESTED ACCESSORY QUALIFIERS: | |
|---|---|
| ABUNDANT | 15%-20% |
| MODERATELY ABUNDANT | 10%-15% |
| MINOR | 5%-10% |
| RARE | 1%- 5% |
| TRACE | <1% |

| SUGGESTED POROSITY QUALIFIERS: | |
|---|---|
| VERY GOOD | >20% |
| GOOD | 15%-20% |
| FAIR | 10%-15% |
| POOR | 5%-10% |
| TRACE | <5% |

**SUGGESTED CARBONATE POROSITY CLASSIFICATION**

| INTERGRANULAR | VUGULAR |
|---|---|
| I. Compact, interlocking crystals or fully cemented grains | A. 20μ |
| II. Chalky texture, grain size 20μ | B. 20μ - 125μ |
| III. Partially cemented grains or dolomite rhombs 20μ e.g. Sucrosic Dolomite | C. 125μ - 2000μ |
| | D. 2000μ - (2m.m) |

**Fig. 11** Percentage estimation chart

Clastic rocks may have intergranular or fracture porosity. Unconsolidated sands are assumed to have good porosity, although original halitic or clay cement may have been washed out of samples collected at the well site.

Carbonate rocks may have intergranular (intercrystalline), secondary (vugular), or fracture porosity. A classification of intergranular and vugular porosity is shown in Fig. 11.

The relationship of porosity to permeability should be noted, e.g., interconnected vugular porosity.

### Hydrocarbon Indications

A full treatment of well-site hydrocarbon evaluation, including hydrocarbon indications in samples, is given in section 4.2.

### Other Features

In this section, any other relevant information should be included.

—Fossils, especially diagnostic fossils
—Concretions and nodules
—Stratification
—Tectonic and structural indicators
—Unconformities
—Sedimentary features
—Extent of diagenesis or metamorphism
—Mineral replacement

### Artificially Introduced Materials

Any artificially introduced materials should be noted to help in future sample investigation.

—Lost-circulation material such as mica
—Mud conditioners such as gypsum
—Hydrocarbons such as pipe dope or diesel in the mud

## 4.2 Hydrocarbon Show Evaluation

### 4.21 Preparation

Many points should be checked constantly, starting long before forecast entry into prospective zones.

Gas collection and recording systems must be checked to see if they are operative and calibrated.

Lag times must be correct. Calculate the theoretical lag time and compare it with a carbide test. Do not run the carbide test, however, anywhere near the prospective zones.

Check that shaker screens are being used and that they are not allowing small particles through the screen. If that is the case, check the use and operation of the desanders and desilters.

Try to ensure that as constant a circulation as possible is being used and that a change of mud pits and systems will not be carried out near prospective zones.

Check that the mud loggers' ancillary equipment, such as pump stroke counters, pit level, mud weight and ROP, are in good working order. Find out where drilling instrumentation is located in the event of failure of mud-logging instrumentation.

Check that a representative and uncontaminated sample is being collected.

Ensure that enough sample catchers are available. Drilling rates may greatly increase in prospective zones.

Check with the driller that he will notify you of drilling and driller's breaks.

Ensure that drillers and mud loggers know their reporting procedures and responsibilities on hydrocarbon shows and drilling breaks.

Assemble all possible artificially introducable materials that may be mistaken for hydrocarbon shows.

Assemble local crudes for comparison.

## 4.22 Awareness of Misleading Information

Artificially introduced materials may be mistaken for oil shows. These include pipe dopes, greases, and diesel or organic chemicals.

The carbide test produces acetylene. Check that such a test has not been carried out.

Check for mineral fluorescence. Minerals such as gilsonite and calcite will fluoresce under ultraviolet light.

Be aware of the effect of the drilling rate on hydrocarbon shows. The faster the drilling rate, the greater will be the shows and gas peaks.

The higher the pump rate, the greater will be the spread of shows and gas peaks.

Mud overbalance will reduce or eliminate hydrocarbon shows.

Be aware of differential lag times between cuttings, oil, and gas. These materials will travel up the annulus at different rates, depending on mud viscosity.

High mud weight and viscosity will reduce gas readings since the separation rate of the gas from the mud will be reduced.

Be aware of the depth of the origin of gas shows. A small gas bubble at a great depth will increase the apparent gas show.

The composition of formation fluids and gases should be considered. Air introduced at connections may cause mud frothing in the flow line. Gases may be hydrocarbon or may include hydrogen sulfide ($H_2S$), carbon dioxide ($CO_2$), nitrogen ($N_2$), and hydrogen ($H_2$) principally with other minor constituents. Noncombustible gases such as nitrogen and carbon dioxide may depress the effect of hydrocarbons if the sensor used to detect them relies on the combustible properties of hydrocarbons. Hydrogen sulfide is particularly nasty for a number of reasons. It is very poisonous and, after initial exposure, a person's sense of smell is destroyed. Steel is rendered brittle by its presence. $H_2S$ expansion occurs at low pressure; thus, little warning of such a kick is given.

Connection and trip gases should not be mistaken for true shows. They may, however, indicate actual lag times, pipe washout, and mud imbalance.

Mud additives may degrade in hot wells to produce carbon dioxide, hydrogen sulfide, and hydrogen. Thermocatalytic action of the bit on organic material in the formation or mud may produce hydrocarbons.

An imbalance may be encountered in areas of shale diapirism, and this may give rise to gas-cut mud. This situation may be unrelated to present occurrences of abnormal high pressures.

Faults may act as migration paths for formation fluids, possibly causing localized fluid flow into the borehole.

Diagenesis of volcanic beds may give rise to clay minerals, methane, and carbon dioxide. The latter may also be derived from calcareous formations.

The cap rock will often contain significant amounts of gas, which will increase near the reservoir.

A false show may be encountered in water-bearing formations where low salinities allow for solution of large amounts of methane, which may be released on drilling.

Coals and lignites may give rise to high gas readings.

The ratio between the various gaseous hydrocarbons can provide some useful information about potential reservoirs (see Fig. 12).

## 4.23 Background to Procedure

The following procedure is for evaluation of shows under rock bit, diamond bit, or coring conditions, including side-wall cores and samples from bits and stabilizers. Appendix III contains a hydrocarbon show evaluation report form that may be referred to at this time.

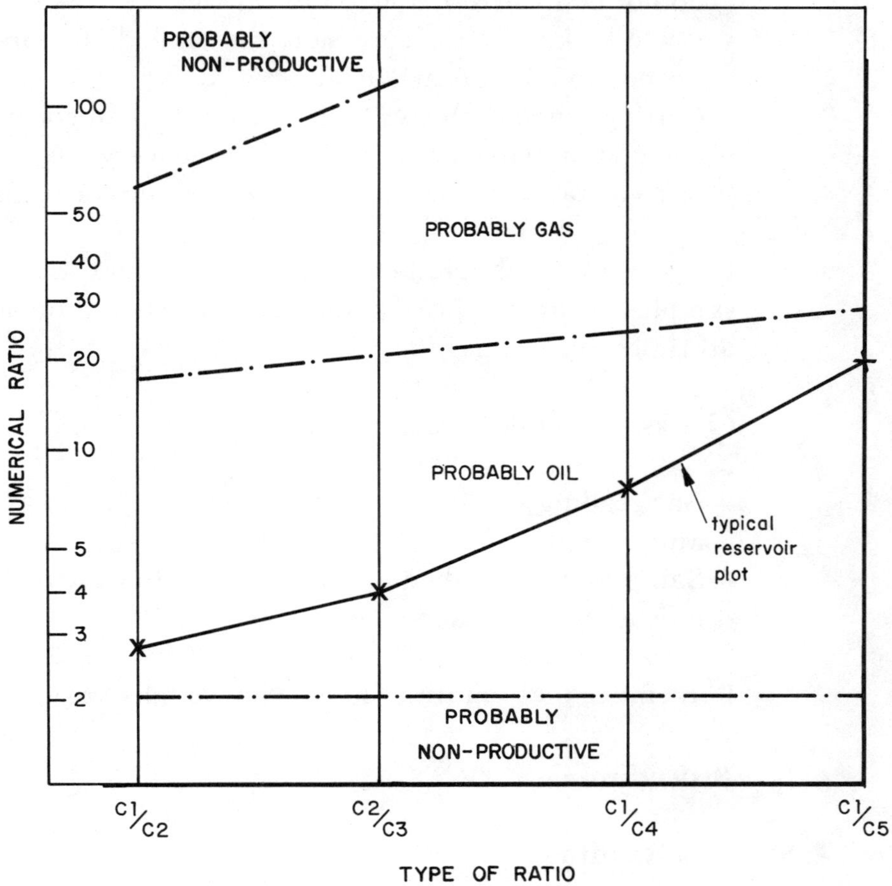


**Fig. 12** Empirical reservoir indicator

—Diamond bit (rock bit or coring) cuttings will be very fine; do not be misled by the many cavings
—When looking at the core, note the immediate appearance for possible hydrocarbon signs
—Cuttings should always be taken, even when coring, for correlation when only partial core recovery has occurred or as a standby when core recovery is zero or negligible

Particular care should be taken to collect, and when collecting, samples at drilling breaks. Fluorescence should be noted at all times.

Checks should be made on the following:

—Gas readings
—Mud samples
—Samples of unwashed and washed cuttings, wiped but unwashed cores, and side-wall cores

Remember that lag times for the above may vary.

## 4.24 Procedure

### 4.241 Gas Readings

The interpretation of flow-line gas at the surface may give rise to much error. Gas recording and analysis may give some indication of hydrocarbon contact, the type of reservoir, and the likelihood of production.

### 4.242 Mud Analysis

The following checks should be carried out on mud samples:

—Observe mud for irridescence, fluorescence, and oil and gas pops
—Note the smell of the mud
—Blend and analyze the gas. This is especially important if the mud is heavy and/or viscous.

### 4.243 Sample Analysis

The following procedure and order of procedure is recommended when analyzing a sample:

### Unwashed Sample Under Natural and UV Light

Unwashed cuttings should be placed in water and examined for hydrocarbons. If there is no fluorescence, the sample and water should be agitated and observation for fluorescence should be repeated. Artificially introduced materials tend to rise slowly to the surface and do not spread out, as do natural hydrocarbons. The speed of bleeding and the color and intensity under natural and ultraviolet light should be noted.

### Lightly Washed Wet Sample

The sample should not be dried because light oils and condensates may evaporate.

#### UNDER UV LIGHT

A differentiation should be made between mineral fluorescence, natural hydrocarbon fluorescence, and artificial hydrocarbon fluorescence. Natural hydrocarbon fluorescence will usually be lithology specific, while artificial hydrocarbon fluorescence will be associated with multiple lithologies. A note of percentage of cuttings exhibiting natural hydrocarbon fluorescence and the color and intensity should be made.

#### UNDER ORDINARY LIGHT

The stain of oil emanating from the sample may be visible. It should be noted for its volume and its intensity, since the latter above may be misleading. Stain volume may be related to permeability, although flushing the borehole will reduce stain volume.

Light oils are more prone to exhibit irridescence, while dark stains tend to indicate heavier crudes. Be sure that the observed hydrocarbon is natural.

### Solvent Cut

Chemicals that are commonly used to extract natural hydrocarbons from the sample include chlorothene, trichloroethene, ether, and acetone. Chlorothene is recommended for its inflammability and low toxicity.

The test is carried out microscopically with small samples in a spot plate and in a test-tube with larger (or a number of small) samples. The procedures are as follows:

**SPOT PLATE EXAMINATION**

The small sample is placed in a spot plate, the solvent is introduced, and the color, intensity, and rate of cut are observed in ordinary and ultraviolet light. The sample is crushed and the test is repeated. If there is no cut at this stage, any fluorescence is probably mineral derived.

Generally, the heavier the oil, the greater the cut; however, asphaltic oils show a greater cut than paraffinic oils of the same gravity.

The chromatography of the oil should be noted. A drop sample of hydrocarbon-contaminated solvent should be placed on filter paper and observed under ordinary and ultraviolet light. The characteristics of any separation should be noted, including the number of rings formed.

**TEST-TUBE EXAMINATION**

Larger samples, or a number of smaller samples, may be tested in a similar manner as in a spot plate; the sample should be crushed and the solvent introduced. A standard method should be employed so comparisons can be made with nearby wells, i.e., 1 cc of crushed cuttings in 2 cc of solvent in a standard 10-cc test-tube.

A control test-tube of 3 cc of solvent should also be set up, and a comparison against white paper is then observed. Again, a drop of hydrocarbon-contaminated solvent should be tested for chromatography as before.

## Blender

Place an unwashed sample in a blender, add water, and analyze the emanating gases, preferably chromatographically.

## HCl Oil Reaction

Hydrocarbons present in a sample (either natural or artificial) will cause carbon dioxide bubbles, released by the action of hydrochloric acid on carbonates, to enlarge by forming an oil film around the bubbles. Thus, the reaction is more prone to frothing. This test is very sensitive, but it does not differentiate between natural or artificial hydrocarbons.

## 4.3 Casing Point Determination

The well-site geologist may be required to input lithological and/or pressure regime information into the casing-point decision-making. It is advisable to inform the mud-logging staff of any marker horizons that may be encountered (and require them to report on such) as well as any pointers that will help in evaluating the pressure regime.

## 4.4 Coring Point Determination

### 4.41 General

The cost of rig time is such that a decision to core is often issued by shore-based personnel. TD cores are often taken as a matter of policy.

Cores may be taken for a number of reasons, principally:

—To determine reservoir parameters
—To calibrate wire-line logs
—To determine hydrocarbon contents
—To establish stratigraphy

### 4.42 Preparation

In anticipation of coring, the following should be checked:

—Sample, gas collection and recording, and ROP systems are operative. Note that the ROP should tide independent or corrections should be applied.
—Drilling program requirements are understood by the drilling and mud-logging staff, e.g., marker horizons or drilling breaks should be notified

### 4.43 Procedure

During coring, a detailed analysis of cuttings should be made. In diamond drilling, the cuttings will be very small and large cavings should be ignored. Analysis of cuttings is required to aid correlation in the case of partial recovery and as a standby in the case of zero recovery. Samples should be taken from the desander and desilter.

Liaison with the coring engineer and/or driller is important. There should be a two-way flow of information: The well-site geologist may be in a position to advise on lithology, and the driller may know what is being drilled by looking at the ROP, pump pressure, and torque.

It is useful and advisable to plot formation-dependent parameters such as ROP, pump pressure, and torque against cuttings lithology, noting that these changes may be very small. Wire-line log information may be added subsequently. Appendix IV contains sample core log headings. The retrieval and packing of the core is set out in section 3.5.

## 4.5 Side-Wall Core Point Determination

### 4.51 General

The well-site geologist generally will be responsible for locating side-wall core points. Obviously, policy guidance from shore should already have been obtained.

Side-wall cores may be taken for the following reasons:

—Paleontological (stratigraphic) evaluation
—Exact lithological identification
—Determination of maturation
—Evaluation of hydrocarbon content

### 4.52 Procedure

—Assess side-wall core points from ROP, gamma/sonic, and lithology logs
—Give depths and expected lithologies to the engineer so he can determine the optimum configuration of bullet, charge, and wire
—Check the depth during firing
—Ensure that a correct depth/shooting sequence is followed so samples are identified correctly
—Be sure that labeling, packaging, examination, and description are carried out adequately

## 4.6 Velocity Shot Point Determination

The well-site geologist may be more or less involved in velocity shot point determination. At the least, he will be required to provide geological tops. A geophysicist may be present to plan or supervise the survey.

### 4.61 Information Required

—Casing points
—Sonic logs
—Caliper logs
—Deviation surveys
—Geological tops

### 4.62 Effect of Local Conditions

—No shots can be taken near casing joints
—Where cycle skipping on the sonic log, shots should be taken above and below this area
—Where sonic log runs join, ensure a good coverage
—Observe caliper log and deviation surveys to ensure the phone is clamped to a good hole wall

As soon as velocity shot points have been worked out, they should be given to the engineer.

# 5. LIAISON AND COOPERATION

Liaison and cooperation are essential to the smooth running of the drilling operation. Unfortunately, the drillers' and accountants' desires to complete the hole as quickly as possible usually conflict with the geologists' quest for information.

Liaison with well-site geologist and shore-based operators and contractors' personnel is a primary responsibility of the well-site geologist. Relationships on the rig should be based on a two-way exchange basis. The well-site geologist has geological information at hand that may aid drillers and mud engineers to help them carry out their jobs more efficiently and cheaply. Likewise, contractors' personnel, especially drillers, very often have an excellent feel and experience of the lithology being drilled.

It is important for the well-site geologist to familiarize himself with the various rig operations since they help him carry out his job better, enable him to talk to rig personnel on their own terms, and show rig personnel that he takes an interest in operations outside his own limited sphere.

# 6. DISPATCH AND NOTIFICATION

The well-site geologist normally is responsible for the dispatch and notification of samples and logs to the shore base. Samples include:

—Cuttings
—Cores, including side-wall cores
—Formation fluids

Logs include (excluding regular reports):

—Mud log, pressure log, and data acquisition log
—Lithological log
—Wire-line logs
—Core logs, including side-wall core logs

The following points should be kept in mind:

—Samples should be as secure as possible
—Relevant personnel should be aware of commercial confidentiality and the importance of samples and logs
—The shore base should be notified of sample and log details, including a list, mode of transport, and expected time and place of arrival
—A duplicate set of logs should be retained so that losing one set does not mean a complete loss of information

# 7. SECURITY

The well-site geologist should make himself responsible for the security of information at the well site. He should ensure that confidential information does not become common knowledge on the rig, and that confidential information is as inaccessible as possible to any person that may be actively seeking such.

The following precautions should be taken:

—Only authorized personnel should be allowed in the mud-logging cabin
—Only authorized personnel should be allowed in the wire-line logging cabin
—During core retrieval, personnel not included in the operation should not be present. For examination and packaging, cores should be taken to a private place.

The well-site geologist should keep secure any information brought to the well or gained at the well. He should ensure complete retrieval at the end of the well.

It is suggested that the operator give a good amount of information about the well to the contract well-site geologist and that this situation holds for the mud-logging personnel. Thus, these contractors will be better informed and more willing and able to provide a better service.

# APPENDIX I CONVERSION TABLES

## Length

| TO / FROM | 1 cm | 1 m | 1 km | 1 in. | 1 ft | 1 yd | 1 mi. (stat) | 1 mi. (naut) |
|---|---|---|---|---|---|---|---|---|
| 1 cm | 1 | 0.01 | $1 \times 10^{-5}$ | 0.3937 | 0.0328 | 0.0109 | $621 \times 10^{-8}$ | $539 \times 10^{-8}$ |
| 1 m | 100 | 1 | 0.001 | 39.37 | 3.281 | 1.094 | $621 \times 10^{-6}$ | $539 \times 10^{-6}$ |
| 1 km | $1 \times 10^{5}$ | 1,000 | 1 | 39,370 | 3,281 | 1,094 | 0.621 | 0.539 |
| 1 in. | 2.54 | 0.0254 | $254 \times 10^{-7}$ | 1 | 0.0833 | 0.0278 | $158 \times 10^{-7}$ | $137 \times 10^{-7}$ |
| 1 ft | 30.48 | 0.3048 | $305 \times 10^{-6}$ | 12 | 1 | 0.333 | $164 \times 10^{-6}$ | $189 \times 10^{-6}$ |
| 1 yd | 91.44 | 0.9144 | $914 \times 10^{-6}$ | 36 | 3 | 1 | $568 \times 10^{-6}$ | $493 \times 10^{-6}$ |
| 1 mi. (stat) | 160,984 | 1,609 | 1.609 | 63,360 | 5,280 | 1,760 | 1 | 0.868 |
| 1 mi. (naut) | 185,322 | 1,853 | 1.853 | 72,961 | 6,080 | 2,027 | 1.152 | 1 |

## Pressure

| TO / FROM | 1 dyne/cm² | 1 g/cm² | 1 kg/cm² | 1 Newton/cm² | 1 atm | 1 psi |
|---|---|---|---|---|---|---|
| 1 dyne/cm² | 1 | $102 \times 10^{-5}$ | $102 \times 10^{-8}$ | 0.1 | $987 \times 10^{-9}$ | $145 \times 10^{-7}$ |
| 1 g/cm² | 980.7 | 1 | $1 \times 10^{-3}$ | 98.07 | $968 \times 10^{-6}$ | $142 \times 10^{-4}$ |
| 1 kg/cm² | 980,700 | 1,000 | 1 | 98,070 | 0.968 | 14.22 |
| 1 Newton/cm² | 10 | 0.010 | $102 \times 10^{-7}$ | 1 | $987 \times 10^{-8}$ | $145 \times 10^{-6}$ |
| 1 atm | $101 \times 10^{4}$ | 1,033 | 1.033 | 101,300 | 1 | 14.7 |
| 1 psi | 68,950 | 70.31 | 0.073 | 6,895 | $681 \times 10^{-4}$ | 1 |

## Density

| TO / FROM | g/cm³ | kg/m³ | lb/in.³ | lb/ft³ | lb/gal (US) |
|---|---|---|---|---|---|
| 1 g/cm³ | 1 | 1,000 | 0.0361 | 62.43 | 8.345 |
| 1 kg/m³ | $1 \times 10^{-3}$ | 1 | $361 \times 10^{-7}$ | 0.0624 | $835 \times 10^{-5}$ |
| 1 lb/in.³ | 27.68 | 27,680 | 1 | 1,728 | 231 |
| 1 lb/ft³ | 0.0160 | 16.02 | $579 \times 10^{-6}$ | 1 | 0.1337 |
| 1 lb/gal (US) | 0.1198 | 119.8 | $433 \times 10^{-5}$ | 7.481 | 1 |

## Weight

| TO / FROM | 1 g | 1 kg | 1 oz | 1 lb | 1 ton (metric) | 1 ton (long) | 1 ton (short) |
|---|---|---|---|---|---|---|---|
| **1 g** | 1 | $1 \times 10^{-3}$ | 0.035 | $220 \times 10^{-5}$ | $1 \times 10^{-6}$ | $984 \times 10^{-9}$ | $110 \times 10^{-8}$ |
| **1 kg** | 1,000 | 1 | 35.27 | 2.205 | $1 \times 10^{-3}$ | $984 \times 10^{-6}$ | $110 \times 10^{-5}$ |
| **1 oz** | 28.35 | $283 \times 10^{-4}$ | 1 | $625 \times 10^{-4}$ | $284 \times 10^{-7}$ | $279 \times 10^{-7}$ | $313 \times 10^{-7}$ |
| **1 lb** | 453.6 | 0.453 | 16 | 1 | $453 \times 10^{-6}$ | $446 \times 10^{-6}$ | $500 \times 10^{-6}$ |
| **1 ton (metric)** | $1 \times 10^{6}$ | 1,000 | 35,270 | 2,205 | 1 | 0.984 | 1.102 |
| **1 ton (long)** | $102 \times 10^{4}$ | 1,016 | 35,840 | 2,240 | 1,016 | 1 | 1.120 |
| **1 ton (short)** | $907 \times 10^{3}$ | 907.2 | 32,000 | 2,000 | 0.907 | 0.893 | 1 |

## Volume

| TO / FROM | 1 $cm^3$ | 1 $m^3$ | 1 $in.^3$ | 1 $ft^3$ | 1 gal (Imp) | 1 gal (US) | 1 bbl | 1 acre-ft |
|---|---|---|---|---|---|---|---|---|
| **1 $cm^3$** | 1 | $1 \times 10^{-6}$ | 0.0610 | $353 \times 10^{-7}$ | $220 \times 10^{-6}$ | $264 \times 10^{-6}$ | $629 \times 10^{-8}$ | $811 \times 10^{-12}$ |
| **1 $m^3$** | $1 \times 10^{6}$ | 1 | 61,023 | 35.31 | 220 | 264.2 | 6.29 | $811 \times 10^{-6}$ |
| **1 $in.^3$** | 16.39 | $164 \times 10^{-7}$ | 1 | $579 \times 10^{-6}$ | $361 \times 10^{-5}$ | $433 \times 10^{-5}$ | $103 \times 10^{-6}$ | $133 \times 10^{-10}$ |
| **1 $ft^3$** | 28,316 | 0.028 | 1,728 | 1 | 6.229 | 7.481 | 0.178 | $229 \times 10^{-7}$ |
| **1 gal (imp.)** | 4,546 | $454 \times 10^{-5}$ | 227.4 | 0.161 | 1 | 1.201 | 0.029 | $369 \times 10^{-8}$ |
| **1 gal (US)** | 3,785 | $378 \times 10^{-5}$ | 231 | 0.134 | 0.833 | 1 | 0.024 | $307 \times 10^{-8}$ |
| **1 bbl** | 159,000 | 0.159 | 9,702 | 5.615 | 34.97 | 42 | 1 | $129 \times 10^{-6}$ |
| **1 acre-ft** | $123 \times 10^{7}$ | $1233 \times 10^{3}$ | $753 \times 10^{5}$ | 43,560 | 271,300 | 325,900 | 7,785 | 1 |

## Area

| TO / FROM | 1 $cm^2$ | 1 $m^2$ | 1 hectare | 1 $in.^2$ | 1 $ft^2$ | 1 $yd^2$ | 1 acre | 1 $mi.^2$ (stat) |
|---|---|---|---|---|---|---|---|---|
| **1 $cm^2$** | 1 | $1 \times 10^{-4}$ | $1 \times 10^{-8}$ | 0.155 | $108 \times 10^{-5}$ | $120 \times 10^{-6}$ | $247 \times 10^{-10}$ | $386 \times 10^{-13}$ |
| **1 $m^2$** | 10,000 | 1 | $1 \times 10^{-4}$ | 1,550 | 10.76 | 1.196 | $247 \times 10^{-6}$ | $386 \times 10^{-9}$ |
| **1 hectare** | $1 \times 10^{8}$ | 10,000 | 1 | $155 \times 10^{5}$ | $108 \times 10^{3}$ | 11,960 | 2.471 | $386 \times 10^{-5}$ |
| **1 $in.^2$** | 6.452 | $645 \times 10^{-6}$ | $645 \times 10^{-10}$ | 1 | $694 \times 10^{-5}$ | $772 \times 10^{-6}$ | $159 \times 10^{-9}$ | $249 \times 10^{-12}$ |
| **1 $ft^2$** | 929 | 0.0929 | $929 \times 10^{-8}$ | 144 | 1 | 0.1111 | $230 \times 10^{-7}$ | $359 \times 10^{-10}$ |
| **1 $yd^2$** | 8,361 | 0.8361 | $836 \times 10^{-7}$ | 1,296 | 9 | 1 | $207 \times 10^{-6}$ | $323 \times 10^{-9}$ |
| **1 acre** | $405 \times 10^{5}$ | 4,047 | 0.4047 | $627 \times 10^{4}$ | 43,560 | 4,840 | 1 | $156 \times 10^{-5}$ |
| **1 mi. (stat)** | $259 \times 10^{8}$ | $259 \times 10^{4}$ | 259 | $401 \times 10^{7}$ | $279 \times 10^{5}$ | $310 \times 10^{4}$ | 640 | 1 |

## Temperature

Degrees Centigrade (°C) = 5/9 × (°F − 32)
Degrees Fahrenheit (°F) = (9/5 × °C) + 32

## Specific Gravity

°API = (141.5/SG) − 131.5
SG = 141.5/(°API + 131.5)

# APPENDIX II GENERAL ABBREVIATIONS

| | |
|---|---|
| barrel | bbl |
| barrels of oil per day | BOPD |
| barrels per day | b/d |
| billion cubic feet (billion = $10^9$) | Bcf |
| billion standard cubic feet | Bscf |
| bottom-hole temperature | BHT |
| bottom-hole pressure | BHP |
| casing | csg |
| connection gas | CG |
| core bit | CB |
| degree API | °API |
| degree centrigrade | °C |
| degree Fahrenheit | °F |
| degree Kelvin | °K |
| d exponent | dx |
| d exponent corrected | dxc |
| density | $\rho$ |
| drill-stem test | DST |
| effective circulating density | ECD |
| filter cake | FC |
| gallons per minute | gal/min (gpm) |
| gas-oil ratio | GOR |
| inside diameter | ID |
| kelly bushing | KB |
| millidarcy | md |
| million cubic feet | MMcf |
| mud weight | MW |
| new bit | NB |
| no returns | NR |
| outside diameter | OD |
| parts per million | ppm |
| porosity | $\phi$ |
| pounds per cubic foot | lb/cu ft |
| pounds per gallon | lb/gal (ppg) |
| pounds per square inch | psi |
| pounds per square inch absolute | psia |
| pounds per square inch gauge | psig |
| rate of penetration | ROP |
| rerun bit | RRB |
| revolutions per minute | RPM |
| rotary table | RT |
| salinity | S |
| sea level | SL |
| specific gravity | sg |
| strokes per minute | spm |
| total depth | TD |
| trip gas | TG |
| viscosity | V |
| water depth | wd |
| weight on bit | WOB (W) |

# LITHOLOGICAL ABBREVIATIONS

| | |
|---|---|
| about | abt |
| above | abv |
| abundant | abnt |
| acicular | acic |
| agglomerate | aglm |
| aggregate | agg |
| algae, algal | Alg |
| altered, altering | alt |
| amorphous | amor |
| amount | amt |
| angular | ang |
| anhedral | anhed |
| anhydrite, anhydritic | anhy |
| apparent | apr |
| appears | aprs |
| approximate, approximately | approx |
| aragonite | arag |
| arenaceous | aren |
| argillaceous | arg |
| arkose, arkosic | ark |
| asphalt, asphaltic | asph |
| at | @ |
| average | av |
| band, banded | bnd |
| barite, baritic | bar |
| basalt | bas |
| bed | bd |
| bedded | bdd |
| bedding | bdg |
| bentonite, bentonitic | bent |
| biotite | biot |

Compiled by the AAPG Committee on Stratigraphic Correlations. Plurals of nouns may be indicated by adding *s* to the abbreviation.

| | |
|---|---|
| bitumen, bituminous | bit |
| black | blk |
| block, blocky | blky |
| blue, bluish | bl |
| botryoidal | btry |
| boulder | bldr |
| Bracchiopod | Brac |
| breccia, brecciated | brec |
| bright | bri |
| brittle | brit |
| brown | brn |
| bryozoa | Bry |
| calcite, calcareous | calc |
| carbonaceous | carb |
| cavernous | cav |
| caving | cvg |
| cement, cemented | cmt |
| center, centered | cntr |
| Cephalopod | Ceph |
| chalcedony | chal |
| chalk, chalky | chk |
| chert | cht |
| cherty | chty |
| chitin, chitinous | chit |
| clastic | clas |
| clay, clayey | cly |
| claystone | clyst |
| clean | cln |
| clear | clr |
| cluster | cls |
| coarse, coarsely | c |
| cobble | cbl |
| color, colored | col |
| common | com |

| | |
|---|---|
| compact | cpct |
| concentric | cncn |
| conchoidal | conch |
| concretion, concretionary | conc |
| conglomerate | cgl |
| conodont | Cono |
| contact | ctc |
| contorted | cntrt |
| coquina | coq |
| covered | cov |
| crenulated | cren |
| crevice | crev |
| crinkled | crnk |
| crinoid, crinoidal | Crin |
| crossbedded, crossbedding | xbd, xbdg |
| cross-laminated | xlam |
| cross-stratified | xstrat |
| cryptocrystalline | crpxl |
| cryptograined | crpgr |
| crystal, crystalline | xl |
| cuttings | ctgs |
| dark | dk |
| dead | dd |
| debris | deb |
| decrease, decreasing | decr |
| dendritic | dend |
| dense | dns |
| determine | dtrm |
| detrital, detritus | dtrl |
| diameter | dia |
| difference | dif |
| disseminated | dism |
| dolocast, dolocastic | dolc |
| dolomite, dolomitic | dol |
| dolomold, dolomoldic | dolmd |
| druse, drusy | drsy |
| earthy | rthy |
| Echinoid | Ech |
| elliptical | elip |
| elongate | elg |
| embedded | embd |
| enlarged | enl |
| equivalent | equiv |
| euhedral | euhed |
| evaporitic | evap |
| expose, exposed, exposure | exp |
| extrusion, extrusive | extr |
| faceted | fac |
| faint | fnt |
| fair | fr |

| | |
|---|---|
| fault | flt |
| fauna | fau |
| feldspar, feldspathic | fld |
| ferruginous | Fe |
| fibrous | fib |
| figured | fig |
| fine, finely | f |
| fissile | fis |
| flaggy | flgy |
| flake | flk |
| flaky | flky |
| flat, flattened | flat |
| floating | fltg |
| fluorescence | flor |
| foliated | fol |
| foraminifera | Foram |
| formation | fm |
| fossil, fossiliferous | fos |
| fracture, fractured | frac |
| fragment, fragmental | frag |
| fresh | frs |
| friable | fri |
| frosted | fros |
| Fusulinid | Fus |
| gabbro | gab |
| Gastropod | Gast |
| glassy | gl |
| glauconite, glauconitic | glau |
| gloss, glossy | glos |
| gneiss | gns |
| good | g |
| grade, grades, graded | grd |
| grading | grdg |
| grain, grained | gr |
| granite | grnt |
| granular | gran |
| granule | grnl |
| graptolite | Grap |
| gravel | gvl |
| gray | gy |
| graywacke | gywke |
| greasy | gsy |
| green | gn |
| gritty | grty |
| gypsum, gypsiferous | gyp |
| hackly | hky |
| hard | hd |
| heavy | hvy |
| hematite, hematitic | hem |
| hexagonal | hex |

| | | | |
|---|---|---|---|
| high | hi | maroon | mar |
| horizontal | hztl | massive | mas |
| hydrocarbon | hydc | material, matter | mat |
| igneous | ig | matrix | mtx |
| imbedded | imbd | maximum | max |
| impression | imp | median | mdn |
| intrusion, intrusive | incl | medium | m |
| increase, increasing | incr | member | mbr |
| indistinct | indst | metamorphic | meta |
| indurated | ind | mica, micaceous | mica |
| interbedded | intbd | microcrystalline | micxl |
| intercalated | intcl | microfossil, microfossiliferous | micfos |
| intercrystalline | intxl | micrograined | micgr |
| interfingered | intfr | micromicaceous | mic-mica |
| intergranular | intgran | middle | mid |
| intergrown | intgwn | mineral, mineralized | mnrl |
| interlaminated | intlam | minimum | min |
| interstitial | intstl | minor | mnr |
| interval | intv | minute | mnut |
| intraformational | intfm | moderate | mod |
| intrusion, intrusive | intr | mollusca | Mol |
| invertebrate | invrtb | mottled, mottling | mot |
| iron | Fe | mudstone | mdst |
| ironstone | Fe-st | muscovite | musc |
| irregular | ireg | nacreous | nac |
| irridescent | irid | no, non- | n. |
| jasper, jasperoid | jasp | nodule | nod |
| jointed | jtd | numerous | num |
| joints | jts | object | obj |
| kaolin | kao | ochre | och |
| laminated | lam | odor | od |
| large, larger | lrg | oil | o |
| lavendar | lav | oil sand | o. sd |
| leached | lchd | oil stain | o. stn |
| ledge | ldg | olive | olv |
| lentil, lenticular | len | oolicast, oolicastic | ooc |
| light, lighter | lt | oolite, oolitic | ool |
| lignite, lignitic | lig | oomold, oomoldic | oom |
| limestone | ls | opaque | op |
| limonite, limnitic | lmn | orange | orng |
| limey | lmy | organic | org |
| lithic, lithology, lithographic | lith | orthoclase | orth |
| little | ltl | Ostracod | Ost |
| long | lg | oxidized | ox |
| loose | lse | part, partly | pt |
| lower | low | parting | ptg |
| lumpy | lmpy | pearl, pearly | prly |
| luster | lstr | pebble | pbl |
| magnetic | magn | pebbly | pbly |
| marlstone | mrlst | pelecypod | plcy |

|  |  |
|---|---|
| pellet | pel |
| permeability | perm |
| petroleum, petroliferous | pet |
| phosphate, phosphatic | phos |
| pink | pk |
| pinpoint | pp |
| pisolite, pisolitic | piso |
| pitted | pit |
| plagioclase | plag |
| plant fossils | pl fos |
| plastic | plas |
| platy | plty |
| polish, polished | pol |
| poor, poorly | p |
| porcelaneous | porc |
| porosity, porous | por |
| possible, possibility | pos |
| predominant, predominantly | pred |
| preserved, preservation | pres |
| primary | prim |
| prism, prismatic | pris |
| probable, probably | prob |
| prominent, prominently | prom |
| pseudo- | psdo |
| purple | purp |
| pyrite, pyritized | pyr |
| pyrobitumen | pyrbit |
| pyroclastic | pyrclas |
| quartz | qtz |
| quartzite | qtzt |
| quartzitic | qtzc |
| quartzose | qtzs |
| radiate, radiating | rad |
| range, ranging | rng |
| rare | rr |
| regular | reg |
| remains, remnant | rmn |
| replaced, replacing, replacement | repl |
| residue, residual | resd |
| resinous | rsns |
| rhomb, rhombic | rhmb |
| rock | rk |
| round, rounded | rd |
| rubbly | rbly |
| sample | spl |
| sand | sd |
| sandstone | ss |
| sandy | sdy |
| saturated, saturation | sat |
| scales | sc |
| scarce | scs |
| scattered | scat |
| schist | sch |
| Scolecodonts | Scol |
| secondary | sec |
| sediment, sedimentary | sed |
| selenite | sel |
| shadow | shad |
| shale | sh |
| shaly | shy |
| siderite, sideritic | sid |
| silica, siliceous | sil |
| silky | slky |
| silt | slt |
| siltstone | sltst |
| silty | slty |
| size | sz |
| slabby | slab |
| slickensided | sks |
| slight, slightly | sl |
| small | s |
| smooth | sm |
| soft | sft |
| solution | sol |
| sort | srt |
| sorted | srtd |
| sorting | srtg |
| speck, speckled | spec |
| sphalerite | sphal |
| spherules | sph |
| spicule, spicular | spic |
| splintery | splty |
| sponge | Spg |
| spore | Spr |
| spot, spotted, spotty | sp |
| stain, stained, staining | stn |
| stippled | stip |
| stone | st |
| strata, stratified, stratification | strat |
| streak | str |
| striated | stri |
| stringer | strg |
| Stromatoporoid | Strom |
| structure | struc |
| stylolite | styl |
| subangular | sbang |
| subhedral | sbhed |
| subrounded | sbrd |
| sucrose | suc |

| | | | |
|---|---|---|---|
| sulfur | S | variegated | vgt |
| surface | surf | varved | vrvd |
| tabular | tab | vein | vn |
| texture | tex | vertebrate | vrtb |
| thick | thk | very | v |
| thin | thn | vesicular | ves |
| throughout | thru | vitreous | vit |
| tight, tightly | tt | volcanics | volc |
| trace | tr | vug, vuggy, vugular | vug |
| translucent | trnsl | water | wtr |
| transparent | trnsp | wavy | wvy |
| Trilobite | Trilo | waxy | wxy |
| tripoli, tripolitic | trip | weather, weathered | wthr, wthrd |
| tubular | tub | well | w |
| tuffaceous | tuf | white | wh |
| unconformity | unconf | with | / |
| unconsolidated | uncons | yellow | yel |
| upper | up | zone | zn |
| variable | var | | |
| varicolored | vcol | | |

# APPENDIX III McKinlay-Smith REPORT SHEETS

## MORNING GEOLOGICAL REPORT

REPORT NO: ____________ PAGE NO: ________

OPERATOR: ____________

WELL: ____________

COUNTRY: ____________

DEPTH at 0600: ____________

FOOTAGE to 0600: ____________

PRESENT OPERATION: ____________

| SUMMARY of OPERATION for 24 HRS. ENDING at 0600 |
|---|
| |
| |
| |

| DEPTHS | R.O.P. | LITHOLOGY, incl. TOPS |
|---|---|---|
| | | |
| | | |
| | | |
| | | |
| | | |
| | | |
| | | |
| | | |
| | | |
| | | |
| | | |
| | | |
| | | |
| | | |
| | | |
| | | |
| | | |
| | | |

| HYDROCARBON SHOWS and MUD GAS ANALYSIS |
|---|
| |
| |
| |
| |
| |

| REMARKS: |
|---|
| |
| |
| |

| GEOLOGIST: | |
|---|---|
| SIGNATURE: | |
| DATE: | |

## WEEKLY GEOLOGICAL REPORT

REPORT NO: ____________ PAGE NO: ________

OPERATOR: ____________

WELL: ____________

COUNTRY: ____________

REPORT INTERVAL: ____________

DATE of REPORT: ____________

| FOR 24 HOURS ENDING AT 0600 ON | SUMMARY OF OPERATIONS |
|---|---|
| | |

| FORMATION | TOPS | R.O.P. | LITHOLOGICAL DESCRIPTION | MUD GAS ANALYSIS | | | | | | | HYDROCARBON SHOW EVALUATION | REMARKS |
|---|---|---|---|---|---|---|---|---|---|---|---|---|
| | | | | TOTAL GAS | PETROLEUM VAPOUR | $C_1$ | $C_2$ | $C_3$ | $C_4$ | $C_5$ | | |
| | | | | | | | | | | | | |

| REMARKS: |
|---|
| |

| GEOLOGIST: | SIGNATURE: | DATE: |
|---|---|---|
| | | |

## LOGGING GEOLOGIST DATA SHEET

REPORT NO: ____________ PAGE NO: ______

OPERATOR: ____________

WELL: ____________

COUNTRY: ____________

REPORT INTERVAL: ____________

FROM: ______ TO: ______

DATE: ____________ LOGGING GEOLOGIST: ____________

| DEPTH | R.O.P. | LAG TIME | SAMPLE CUT | SAMPLE CAUGHT | LITHOLOGY DESCRIPTION | POROSITY | TOTAL GAS | $C_1$ | $C_2$ | $C_3$ | $C_4$ | $C_5$ | HYDROCARBON SHOW | REMARKS |
|---|---|---|---|---|---|---|---|---|---|---|---|---|---|---|
| | | | | | | | | | | | | | | |

# HYDROCARBON SHOW EVALUATION REPORT

REPORT NO: ______________________ PAGE NO: ________

OPERATOR: ______________________

WELL: ______________________

COUNTRY: ______________________

REPORT INTERVAL: ______________________

DATE of REPORT: ______________________

| | | |
|---|---|---|
| DRILLING BREAK at: | BACKGROUND GAS: | MUD PARAMETER CHANGES: |
| FLOW CHECK: | CONNECTION GAS: | |
| OPERATIONAL INFORMATION: | TRIP GAS: | |
| | OTHER GAS: | |

**GAS READINGS AT SHOW**

| LAGGED DEPTH | CONTINUOUS MUD GAS ANALYSIS | | | | | | |
|---|---|---|---|---|---|---|---|
| | T.G. | P.V. | $C_1$ | $C_2$ | $C_3$ | $C_4$ | $C_5$ |
| | | | | | | | |
| | | | | | | | |
| | | | | | | | |
| | | | | | | | |
| | | | | | | | |
| | | | | | | | |
| | | | | | | | |

**INDIVIDUAL MUD SAMPLE TESTS**

| LAGGED DEPTH | BLENDER GAS ANALYSIS | | | | | | | OBSERVATIONS on MUD incl. smell, irridescence, fluorescence, gas "pops" etc. |
|---|---|---|---|---|---|---|---|---|
| | T.G. | P.V. | $C_1$ | $C_2$ | $C_3$ | $C_4$ | $C_5$ | |
| | | | | | | | | |
| | | | | | | | | |
| | | | | | | | | |
| | | | | | | | | |
| | | | | | | | | |
| | | | | | | | | |
| | | | | | | | | |

**LITHOLOGIES PRESENT OVER SHOW INTERVAL**

| LAGGED DEPTH | LITHOLOGY incl. POROPERM |
|---|---|
| | |
| | |
| | |
| | |
| | |
| | |
| | |

| LAGGED DEPTH | UNWASHED SAMPLE UNDER U.V. and NATURAL LIGHT<br>SAMPLE TYPE: ________ | LIGHTLY WASHED SAMPLE UNDER U.V. and NATURAL LIGHT<br>SAMPLE TYPE: ________ | SPOT PLATE EXAMINATION for CUT and CRUSH CUT<br>SOLVENT TYPE: ________ | TEST-TUBE CUT EXAMINATION<br>VOL. of SAMPLE: ________<br>VOL. and TYPE of SOLVENT: ________<br>VOL. of TEST TUBE: ________ | CHROMATOGRAPHY | Hcl OIL TEST |
|---|---|---|---|---|---|---|
| | | | | | | |
| | | | | | | |
| | | | | | | |
| | | | | | | |
| | | | | | | |
| | | | | | | |
| | | | | | | |
| | | | | | | |
| | | | | | | |
| | | | | | | |
| | | | | | | |
| | | | | | | |
| | | | | | | |
| | | | | | | |

| ASSESSMENT of SHOW and REMARKS: |
|---|
| |
| |
| |

| GEOLOGIST: | SIGNATURE: | DATE: |
|---|---|---|
| | | |
| | | |
| | | |

# WELL SITE GEOLOGIST'S PRESSURE REPORT

REPORT NO: ______ PAGE NO: ______

OPERATOR: ______

WELL: ______

COUNTRY: ______

REPORT INTERVAL: ______

DATE of REPORT: ______

PERTINENT DATA SINCE:

| SUMMARY of OPERATIONS | LITHOLOGY | HYDROCARBON SHOWS |
|---|---|---|
| | | |

| DEPTH | LITHOLOGY | Dxc | BULK DENSITY | DIFFERENTIAL TEMPERATURE | FLOWLINE CONDUCTIVITY | SHALE FACTOR | DITCH GAS BACKGROUND | DITCH GAS TRIP | DITCH GAS CONNECTION | BLENDER GAS | | | | COMMENTS incl. BOREHOLE CONDITION |
|---|---|---|---|---|---|---|---|---|---|---|---|---|---|---|
| | | | | | | | | | | | | | | |

| TREND DEVIATION: |
|---|
| |
| ACTION TAKEN: |
| |
| FURTHER ACTION RECOMMENDED: |
| |

| PRESSURE PARAMETERS: |
|---|
| |

| GEOLOGIST: | SIGNATURE: | DATE: |
|---|---|---|
| | | |
| PRESSURE ENGINEER: | SIGNATURE: | DATE: |
| | | |

## WIRE-LINE SURVEY REPORT

REPORT NO: ____________ PAGE NO: ______

OPERATOR: ____________

WELL: ____________

COUNTRY: ____________

REPORT INTERVAL: ____________

DATE of REPORT: ____________

| DEVICES RUN | TIME IN | TIMEOUT | DURATION | REMARKS | | | |
|---|---|---|---|---|---|---|---|
| | | | | | | | |
| | | | | | | | |
| | | | | | | | |
| | | | | | | | |
| | | | | | | | |
| SIDEWALL CORES: | | | | | | | |
| | | | | NO. of RUNS: | TOTAL RECEIVED: | BULLETS LEFT in HOLE: | |
| | | | | TOTAL SHOTS: | % RECOVERY: | NO. of CORES ACCEPTED: | |

| MUD and CIRCULATION DETAILS |
|---|
| |
| |
| |
| |
| TESTS on MUD: |
| |
| |
| |

| TOTAL LOGGING TIME | TOTAL TIME LOST | LOST TIME DUE TO |
|---|---|---|
| | | |
| | | |

| GEOLOGICAL INFORMATION INCLUDING TOPS, CORRELATIONS |
|---|
| |

| QUICK-LOOK METHODS CARRIED OUT AT WELLSITE |
|---|
| |

| REMARKS: |
|---|
| |
| |
| |

| GEOLOGIST: | SIGNATURE: | DATE: |
|---|---|---|
| | | |
| | | |
| | | |

# LOG TRANSMITTAL LIST

SHEET NO: ____________

OPERATOR: ____________

WELL: ____________

COUNTRY: ____________

| TYPE OF LOG | RUN NO. | INTERVAL | VERT. SCALE | LOGGING DATE | DATE OF SHIPPING | MEANS OF TRANSPORT | ADDRESSEE | SIGNATURE and POSITION OF SENDER | LOG ARRIVED (✓) |
|---|---|---|---|---|---|---|---|---|---|
| | | | | | | | | | |
| | | | | | | | | | |
| | | | | | | | | | |
| | | | | | | | | | |
| | | | | | | | | | |
| | | | | | | | | | |
| | | | | | | | | | |
| | | | | | | | | | |
| | | | | | | | | | |
| | | | | | | | | | |
| | | | | | | | | | |
| | | | | | | | | | |
| | | | | | | | | | |
| | | | | | | | | | |
| | | | | | | | | | |
| | | | | | | | | | |
| | | | | | | | | | |
| | | | | | | | | | |
| | | | | | | | | | |
| | | | | | | | | | |
| | | | | | | | | | |
| | | | | | | | | | |
| | | | | | | | | | |
| | | | | | | | | | |
| | | | | | | | | | |

## SAMPLE TRANSMITTAL LIST

SHEET NO: ____________

OPERATOR: ____________

WELL: ____________

COUNTRY: ____________

| TYPE OF SAMPLE | BOX NO. | INTERVAL | NO. OF SAMPLES IN BOX | DATE SAMPLES COLLECTED | DATE OF SHIPPING | MEANS OF TRANSPORT | ADDRESSEE | SIGNATURE and POSITION OF SENDER | SAMPLE ARRIVED (✓) |
|---|---|---|---|---|---|---|---|---|---|
| | | | | | | | | | |
| | | | | | | | | | |
| | | | | | | | | | |
| | | | | | | | | | |
| | | | | | | | | | |
| | | | | | | | | | |
| | | | | | | | | | |
| | | | | | | | | | |
| | | | | | | | | | |
| | | | | | | | | | |
| | | | | | | | | | |
| | | | | | | | | | |
| | | | | | | | | | |
| | | | | | | | | | |
| | | | | | | | | | |
| | | | | | | | | | |
| | | | | | | | | | |
| | | | | | | | | | |
| | | | | | | | | | |
| | | | | | | | | | |
| | | | | | | | | | |
| | | | | | | | | | |
| | | | | | | | | | |
| | | | | | | | | | |
| | | | | | | | | | |

# APPENDIX IV McKinlay-Smith LOG HEADING SHEETS

LITHOLOGY LOG

INITIAL STATUS:

SCALE:

| COUNTRY: | DRILLING COMMENCED | | TOTAL DEPTH | K.B. |
|---|---|---|---|---|
| AREA: | DRILLING ENDED | | | SUBSEA |
| LICENCE: | RIG: | | FORMATION AT T.D. | |
| LOCATION N. / E. | GEOLOGISTS: | | | |

**ABBREVIATIONS & SYMBOLS**

| Symbol | Meaning | Abbreviation | Meaning |
|---|---|---|---|
| | CASING SHOE | D.S. | DIRECTIONAL SURVEY |
| | CORED INTERVAL and RECOVERY | T.G. | TOTAL GAS |
| | WIRELINE LOGS | C.G. | CONNECTION GAS |
| | S.W.C. NOT REC. | M.W. | MUD WEIGHT |
| | S.W.C. REC. | V | VISCOSITY |
| | OIL SHOW | L.C.M. | LOST CIRCULATION MATERIAL |
| | GAS SHOW | N.B. | NEW BIT |
| | F.I.T., RFT. | C.B. | CORE BIT |
| | | R.R.B. | RE-RUN BIT |

**LOCATION**

**LITHOLOGY SYMBOLS**

| | | | |
|---|---|---|---|
| Claystone, Mudstone, Shale (argillaceous) | Silty clay, Mudstone or Shale | Silt, Siltstone (silty) | Glauconitic |
| Sand, Sandstone (sandy) | Gravel, Conglomerate (gravelly, conglomeratic) | Sandy gravel, Sandy conglomerate | Bituminous |
| Dolomite (dolomitic) | Limestone, Chalk (calcareous) | Breccia (brecciated) | Microfossils |
| Dolomitic limestone Calcareous dolomite (calcareous/dolomitic) | Gypsum, Anhydrite (gypsiferous, anhydritic) | Halite (halitic) | Pyritic |
| Coal, Lignite (carbonaceous) | Chert (siliceous, cherty) | Intrusive volcanics | K K Potassic |
| No samples | Metamorphics M M (metamorphosed) | Extrusive volcanics T T Tuffaceous | Plant remains |
| | | | Mc Micaceous |
| | | | Macrofossils |
| | | | Shell fragments |

**COMMENTS**

SCALE: AUTHOR: DRAWN: DATE:

| STRAT. | DRILLING RATE SLOW FAST | CUTTINGS LITHOLOGY | DEPTH ft m | INTERPRETED LITHOLOGY | LITHOLOGICAL & HYDROCARBON EVALUATION | HYDROCARBON SHOWS GOOD MOD POOR | MUD GAS ANALYSIS Total Gas Petroleum Vapours $C_1$ $C_2$ $C_3$ $C_4$ $C_5$ |
|---|---|---|---|---|---|---|---|

FINAL CORE LOG

SCALE:

| COUNTRY: | FORMATIONS ENCOUNTERED: | CORE DRILLED | DRILLED DEPTH TO TOP OF CORE | K B |
|---|---|---|---|---|
| AREA: | | CORE RECOVERED | | SUBSEA |
| LICENCE: | | % RECOVERY | DRILLED DEPTH TO BASE OF CORE | K B |
| | | DATE OF CORING | | SUBSEA |

## LITHOLOGY

| | | |
|---|---|---|
| CLAYSTONE, MUDSTONE, SHALE | SILTY CLAY, MUDSTONE or SHALE | SILT, SILTSTONE |
| (ARGILLACEOUS) | | (SILTY) |
| SAND, SANDSTONE | GRAVEL, CONGLOMERATE | SANDY GRAVEL, SANDY CONGLOMERATE |
| (SANDY) | (GRAVELLY, CONGLOMERATIC) | |
| DOLOMITE | LIMESTONE, CHALK | BRECCIA |
| (DOLOMITIC) | (CALCAREOUS) | (BRECCIATED) |
| DOLOMITIC LIMESTONE, CALCAREOUS DOLOMITE | GYPSUM, ANHYDRITE | HALITE |
| (CALCAREOUS, DOLOMITIC) | (GYPSIFEROUS, ANHYDRITIC) | (HALITIC) |
| COAL, LIGNITE | CHERT | INTRUSIVE VOLCANICS |
| (CARBONACEOUS) | (SILICEOUS, CHERTY) | |
| NO SAMPLES | METAMORPHICS | EXTRUSIVE VOLCANICS |
| GLAUCONITIC | METAMORPHOSED | TUFFACEOUS |
| BITUMINOUS | PYRITIC | MICACEOUS |
| MICROFOSSILS | POTASSIC | MACROFOSSILS |
| | PLANT REMAINS | SHELL FRAGMENTS |

## STRUCTURE

| | | |
|---|---|---|
| MASSIVE NO APPARENT BEDDING | SYMMETRICAL RIPPLES | INJECTION STRUCTURE |
| SLIGHTLY BEDDED | INTERFERENCE RIPPLES | SEDIMENTARY DRAGFOLD |
| HORIZONTAL BEDDING | ASSYMETRICAL RIPPLES | PULL-APART |
| CROSS BEDDING | MUD CRACKS | FLAME STRUCTURE |
| LAMINATED | SALT MOULDS | SLUMPS |
| CROSS LAMINATED | TRAILS | MOTTLED, BIOTURBATED |
| FLASER BEDDING | FLUTE CASTS | BURROWED |
| DISTURBED BEDDING | STRIATION CASTS | BOUDINAGE |
| LENTICULAR BEDDING | LOAD CASTS | STYLOLITES |
| WAVY BEDDING | CONE IN CONE | BORINGS, TUBES |
| GRADED BEDDING | COLLAPSE BRECCIA | STROMATOLITE |
| CRINKLED BEDDING | PROD CASTS, BOUNCE CASTS | PLANT REMAINS |
| | | SEAT EARTH, ROOTLETS |

| BED BOUNDARIES | OTHER STRUCTURES | REFERENCE FEATURES |
|---|---|---|
| TRANSITIONAL | FRACTURES | P1 PHOTOGRAPH |
| ABRUPT | MICROFRACTURES | THIN SECTION |
| EROSIONAL | MICROFAULTS | CORE PLUG |
| DISTURBED | LENS | CA CORE ANALYSIS |
| | VUGGY | CORE MISSING / NOT RECOVERED |

## COMMENTS

SCALE AUTHOR DRAWN DATE

| STRAT | DRILLING RATE | GAMMA RAY | DEPTH | LITHOLOGY | STRUCTURES | GRAIN SIZE | SORTING | ROUNDNESS | POROSITY | POROSITY TYPE | PERMEABILITY | LITHOLOGICAL & HYDROCARBON EVALUATION | HYDROCARBON SHOWS | THIN SECTIONS | MUD GAS ANALYSIS | ENVIRONMENT |
|---|---|---|---|---|---|---|---|---|---|---|---|---|---|---|---|---|
| | SLOW FAST | 0 A.P.I. UNITS 100<br>100 200<br>Δt<br>140 μ.secs/ft. 40<br>240 140 | ft m | | | CLASTIC & CARBONATE<br>SILT VF F M C VC | | | EST: −<br>ACT: +<br>CEMENT TYPE & %<br>10 20 30 | | 1 10 100 1000 | | GOOD MOD POOR | | Total Gas<br>Petroleum Vapours<br>$C_1$<br>$C_2$<br>$C_3$<br>$C_4$<br>$C_5$ | |

## SIDE WALL CORE LOG

INITIAL STATUS:

SCALE:

| COUNTRY: | SHOOTING COMMENCED | | CORES FROM: |
|---|---|---|---|
| AREA: | SHOOTING ENDED | | TO: |
| LICENCE: | RIG: | | NO. OF CORES: |
| LOCATION N. | GEOLOGISTS: | | NO. OF RUNS: |
| E. | | | % RECOVERY: |

### LITHOLOGY SYMBOLS

| | | | |
|---|---|---|---|
| Claystone, Mudstone, Shale (argillaceous) | Silty clay, Mudstone or Shale | Silt, Siltstone (silty) | Glauconitic |
| | | | Bituminous |
| Sand, Sandstone (sandy) | Gravel, Conglomerate (gravelly, conglomeratic) | Sandy gravel, Sandy conglomerate | Microfossils |
| Dolomite (dolomitic) | Limestone, Chalk (calcareous) | Breccia (brecciated) | Pyritic |
| | | | K K Potassic |
| Dolomitic limestone Calcareous dolomite (calcareous/dolomitic) | Gypsum, Anhydrite (gypsiferous, anhydritic) | Halite (halitic) | Plant remains |
| | | | Mc Micaceous |
| Coal, Lignite (carbonaceous) | Chert (siliceous, cherty) | Intrusive volcanics | Macrofossils |
| No samples | Metamorphics M M (metamorphosed) | Extrusive volcanics T T (Tuffaceous) | Shell fragments |

### COMMENTS

SCALE: AUTHOR: DRAWN: DATE:

| SHOT NUMBER / RUN NUMBER | STRAT. | DEPTH | LITHOLOGY | GRAIN SIZE CLASTIC & CARBONATE (SLT VF F M C VC) | SORTING | ROUNDNESS | ACCESSORIES | POROSITY EST. ········ ACT. —— / CEMENT TYPE &% / POROSITY TYPE | PERMEABILITY mD 1 10 100 1000 | LITHOLOGICAL & HYDROCARBON EVALUATION | HYDROCARBON SHOWS (GOOD MOD. POOR) | THIN SECTIONS |
|---|---|---|---|---|---|---|---|---|---|---|---|---|
| | | | | | | | | | | | | |

## WELLSITE CORE LOG

INITIAL STATUS

SCALE:

| COUNTRY: | CORING COMMENCED | | CORE DEPTHS | K.B. |
|---|---|---|---|---|
| AREA: | CORING COMPLETED | | | SUBSEA |
| LICENCE: | RIG: | | FORMATIONS ENCOUNTERED: | |
| LOCATION N / E | GEOLOGISTS: | | | |

### LITHOLOGY

CLAYSTONE, MUDSTONE, SHALE — SILTY CLAY, MUDSTONE or SHALE — SILT, SILTSTONE

(ARGILLACEOUS) — (SILTY)

SAND, SANDSTONE — GRAVEL, CONGLOMERATE — SANDY GRAVEL, SANDY CONGLOMERATE

(SANDY) — (GRAVELLY, CONGLOMERATIC)

DOLOMITE — LIMESTONE, CHALK — BRECCIA

(DOLOMITIC) — (CALCAREOUS) — (BRECCIATED)

DOLOMITIC LIMESTONE CALCAREOUS DOLOMITE — GYPSUM, ANHYDRITE — HALITE

(CALCAREOUS, DOLOMITIC) — (GYPSIFEROUS, ANHYDRITIC) — (HALITIC)

COAL, LIGNITE — CHERT — INTRUSIVE VOLCANICS

(CARBONACEOUS) — (SILICEOUS, CHERTY)

NO SAMPLES — METAMORPHICS — EXTRUSIVE VOLCANICS

GLAUCONITIC — METAMORPHOSED — TUFFACEOUS

BITUMINOUS — PYRITIC — MICACEOUS

MICROFOSSILS — POTASSIC — MACROFOSSILS

PLANT REMAINS — SHELL FRAGMENTS

### STRUCTURE

MASSIVE NO APPARENT BEDDING — SYMMETRICAL RIPPLES — INJECTION STRUCT…

SLIGHTLY BEDDED — INTERFERENCE RIPPLES — SEDIMENTARY DRAGFOL…

HORIZONTAL BEDDING — ASSYMETRICAL RIPPLES — PULL-APART

CROSS BEDDING — MUD CRACKS — FLAME STRUCTURE

LAMINATED — SALT MOULDS — SLUMPS

CROSS LAMINATED — TRAILS — MOTTLED, BIOTURBATED

FLASER BEDDING — FLUTE CASTS — BURROWED

DISTURBED BEDDING — STRIATION CASTS — BOUDINAGE

LENTICULAR BEDDING — LOAD CASTS — STYLOLITES

WAVY BEDDING — CONE IN CONE — BORINGS, TUBES

GRADED BEDDING — COLLAPSE BRECCIA — STROMATOLITE

CRINKLED BEDDING — PROD CASTS, BOUNCE CASTS — PLANT REMAINS

SEAT EARTH, ROOTLE…

**BED BOUNDARIES**: TRANSITIONAL — ABRUPT — EROSIONAL — DISTURBED

**OTHER STRUCTURES**: FRACTURES — MICROFRACTURES — MICROFAULTS — LENS — VUGGY

**REFERENCE FEATURES**: P1 PHOTOGRAPH — THIN SECTION — CORE PLUG — CA CORE ANALYSIS — CORE MISSING/ NOT RECOVERE…

### COMMENTS

SCALE — AUTHOR — DRAWN — DATE

| STRAT. | DRILLING RATE (SLOW – FAST) / TORQUE (LOW – HIGH) / PUMP PRESSURE (LOW – HIGH) | DEPTH (ft m) | GRAPHIC LITHOLOGY | STRUCTURE | LITHOLOGICAL & HYDROCARBON EVALUATION | HYDROCARBON SHOWS (GOOD, MOD, POOR) | MUD GAS ANALYSIS: Total Gas; Petroleum Vapours; $C_1$, $C_2$, $C_3$, $C_4$, $C_5$ |
|---|---|---|---|---|---|---|---|

FINAL WELL LOG

SCALE:

INITIAL STATUS:

FINAL STATUS:

| | | | |
|---|---|---|---|
| COUNTRY: | FINAL LOCATION | N. | TOTAL DEPTHS |
| AREA: | | E. | |
| LICENCE: | FORMATION AT T.D: | | |
| SPUD DATE: | | | |
| COMPLETION DATE: | | | |
| TIME TO COMPLETE: | | | |

| TOTAL DEPTHS | | DRILLER | LOGGER |
|---|---|---|---|
| TOTAL MEASURED DEPTH | (a) BELOW K.B. | | |
| | (b) SUBSEA | | |
| TOTAL VERTICAL DEPTH | (a) BELOW K.B. | | |
| | (b) SUBSEA | | |

## LITHOLOGY

Claystone, Mudstone, Shale — (argillaceous)
Silty clay, Mudstone or Shale
Silt, Siltstone — (silty)
Sand, Sandstone — (sandy)
Gravel, Conglomerate — (gravelly, conglomeratic)
Sandy gravel, Sandy conglomerate
Dolomite — (dolomitic)
Limestone, Chalk — (calcareous)
Breccia — (brecciated)
Dolomitic limestone, Calcareous dolomite — (calcareous/dolimitic)
Gypsum, Anhydrite — (gypsiferous, anhydritic)
Halite — (halitic)
Coal, Lignite — (carbonaceous)
Chert — (siliceous, cherty)
Intrusive volcanics
No samples
Metamorphics — (metamorphosed)
Extrusive volcanics — (Tuffaceous)
Glauconitic
Pyritic
Micaceous
Bituminous
Potassic
Macrofossils
Microfossils
Plant remains
Shell fragments

## STRATIGRAPHIC TOP CRITERIA

## LOCATION MAP

## ELEVATION

## ENGINEERING DATA

CASING SHOE
LINER HANGER
LINER SHOE
CEMENT PLUG
BRIDGE PLUG
CORED INTERVAL and RECOVERY
WIRELINE LOGS
S.W.C. RECOVERED
S.W.C. NOT RECOVERED
OIL SHOW
GAS SHOW
F.I.T., R.F.T.
D.S. DIRECTIONAL SURVEY
D.S.T.
PERFORATED INTERVAL

## WIRELINE LOGS

BOTTOM HOLE TEMPERATURE

## CORE INFORMATION

## TESTING INFORMATION

## ABANDONMENT DATA

| BIT SIZE | FROM | TO | CASING SIZE | SHOE AT | CEMENT | FROM | TO | MUD TYPE | FROM | TO |
|---|---|---|---|---|---|---|---|---|---|---|
| | | | | | | | | | | |

## COMMENTS

SCALE: AUTHOR: DRAWN: DATE:

| STRAT. | DRILLING RATE / GAMMA RAY (0–100, 100–200 API UNITS) | DEPTH | LITHOLOGY (GOOD, MOD, POOR) | RESISTIVITY | INTERVAL TRANSIT TIME ΔT (INTEGRATOR: 1ms, 10ms; 140–40, 240–140 µ secs/ft) | DESCRIPTION |
|---|---|---|---|---|---|---|
| | | | | | | |